BIOMASS

ENERGY, POWER, AND ENVIRONMENT

A Series of Reference Books and Textbooks

Editor

PHILIP N. POWERS

Professor Emeritus of Nuclear Engineering
Purdue University
West Lafayette, Indiana

Consulting Editor
Energy Management and Conservation

PROFESSOR WILBUR MEIER, JR.

Head, School of Industrial Engineering
Purdue University
West Lafayette, Indiana

Additional Volumes in Preparation

BIOMASS

Applications, Technology, and Production

NICHOLAS P. CHEREMISINOFF
PAUL N. CHEREMISINOFF
FRED ELLERBUSCH

MARCEL DEKKER, INC.   New York and Basel

Library of Congress Cataloging in Publication Data

Cheremisinoff, Nicholas P.
  Biomass: applications, technology, and production.

  (Energy, power, and environment; v. 5)
  Bibliography: p.
  Includes index.
  1. Biomass energy. I. Cheremisinoff, Paul N.,
joint author. II. Ellerbusch, Fred, joint author.
III. Title. IV. Series.
TP360.C46   662'.669   80-10224
ISBN 0-8247-6933-3

MARCEL DEKKER, INC.
270 Madison Avenue, New York, New York 10016

Current printing (last digit):
10 9 8 7 6 5 4 3 2 1

PRINTED IN THE UNITED STATES OF AMERICA

Preface

In spite of increasing importation of fuels for industry, transportation, and domestic needs, the United States continues to face severe energy problems in both supply and costs. Reserves of oil and gas are diminishing, while expansion of coal and nuclear power remain environmentally controversial issues, to say the least. Once again, as worldwide demand for oil approaches supply capacity, energy costs are rising sharply.

As severe as the problem may appear, solutions are available. There are very large alternative sources of energy and much of the technology to put them to work is known or at hand. One such vast energy resource is biomass. Biomass utilization may be described as a technology that can be put into practice toward a wide variety of energy-recovery applications: it may be small scale, decentralized, energy efficient, and not necessarily capital intensive, using local materials, labor, and ingenuity; or it may extend to large-scale centralized operations. Most importantly, it offers a method of maximizing the use of renewable energy resources. Biomass is a renewable source of energy.

The authors have gone to a wide variety of sources to collect the materials presented in this volume, which is intended as a technical overview of biomass as an energy resource. Primary references have been cited wherever possible for those who may wish to dig deeper and more extensively into the subject.

The authors wish to thank the many individuals who have helped us with their criticism and review of materials presented, as well as the publisher, Marcel Dekker, for his encouragement and assistance.

Nicholas P. Cheremisinoff
Paul N. Cheremisinoff
Fred Ellerbusch

iii

Contents

Contents

1

Biomass and the Earth's Energy Cycle

INTRODUCTION

The human race has set itself apart from the animal kingdom by its intensive thirst for knowledge, its inventiveness and imagination, and its ability
to survive such catastrophic events as war, often accompanied by pestilence
and famine. Today, we are faced with a grave new problem, the energy
crisis, which if unresolved will certainly have dire consequences for civil-
ization in the not too distant future.

By the year 2000 the U.S. gross energy consumption alone is projected
to be 163.4 quadrillion (10^{15}) Btu, more than a twofold increase as compared
to the 1974 U.S. consumption (1). At present, roughly 94% of our total
energy consumption is based upon conventional fossil fuels, i.e., coal,
petroleum, and natural gas. The supplies of natural gas and petroleum are
rapidly approaching depletion, and many of our coal deposits, although
abundant, are environmentally unacceptable because of related air pollution
problems. Nuclear power and oil shale are expected to play an important
role in the energy picture by the year 2000; however, both are also limited
by environmental constraints and so they are not the ultimate solution.
Solar energy is one alternative energy source expected to have some impact
on the problem by slowing down the diminution of supplies of oil and natural
gas. Research, development, and demonstration activities are aimed at
various promising solar energy processes and technologies. Two general
areas are presently being emphasized: direct space heating and electricity
generation. Some suitable technologies are already available, although not
perfected, but solar energy alone is not the answer. Exotic forms of energy
such as wind and geothermal steam are expected to have relatively little
impact on our energy needs in the near future (2,3).

The bioconversion of agricultural products and industrial and municipal
wastes to clean forms of usable fuel is one alternative energy source that

is showing great promise in replacing or at least in part replacing diminishing conventional fuels. In practical terms, biomass is a renewable resource that is capable of supplying nonpolluting, safe fuels. Biomass as a source of energy is not a new idea. For many years the Europeans, for example, have practiced one form of bioconversion by the generation of steam from waste combustion. Today, a worldwide effort is being initiated into research and development of processes that can commercially use wastes, agricultural products, and marine life as forms of energy. To be practical, the energy or fuel produced by bioconversion processes must be storeable, transportable, environmentally acceptable, and within certain economic constraints. This book presents an overview of biomass technology. It examines the state of the art of this fast-growing technology, indicating the directions in which research is now headed and pointing to areas where it should be directed in the future.

THE LIVING WORLD AND THE ENERGY CYCLE

The term biosphere refers to that portion of the earth in which life exists. The terrestrial envelope that constitutes the biosphere is irregular in shape as it is bounded by an indefinite "parabiospheric" region in which various dormant forms of life exist (4). The biosphere is important to life in that it is the region in which water exists in liquid state, enabling plants to grow and utilize energy from the sun through photosynthesis. Furthermore, the interfaces between liquid, solid, and gaseous states of matter are of importance in biological processes.

The sun is the energy source on which all terrestrial life is based. The energy of solar radiation is the driving force of the biological cycle. Only about 0.1% of the energy received by the earth from the sun enters into photosynthetic production of organic matter. This small amount of energy is responsible for the yearly production of thousands of grams of dry organic material per square meter. Roughly 150 to 200 billion tons of dry organic matter is estimated to be produced annually in the world as vegetation in the forests, grasslands, marshes, oceans, estuaries, lakes, rivers, tundras, etc. (5).

Approximately half of the energy tied up in photosynthesis is involved in plant respiration. In land plants, some of this energy is transferred from tissues to a fixed state where it is used or stored.

The major cycles of the biosphere are illustrated in Fig. 1.1. Photosynthetic reduction of carbon dioxide to form organic compounds (carbohydrates) and molecular oxygen is one of the primary biological reactions responsible for life:

$$nCO_2 + 2nH_2O + energy \rightarrow (CH_2O)_n + nO_2 + nH_2O \qquad (1.1)$$

The end products of photosynthesis are atmospheric oxygen and free water in which oxygen is dissolved; in addition, as already noted, it produces

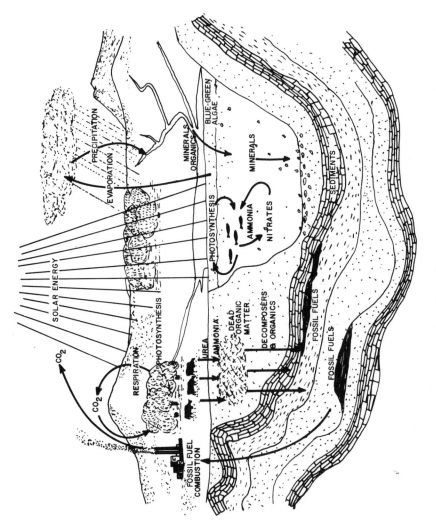

FIG. 1.1 The primary cycles of the biosphere.

organic matter, and the organic decomposition of this matter generates
soils and aquatic sediments. Some of these sediments are buried and subsequently generate organic carbon and fossil fuels, resulting in a loss of
oxygen. Similar oxygen loss occurs through the oxidation of eroding primary
rock and in fossilization. Essential to the operation of the biosphere is the
continual oxidation of the reduced portion, either living or dead, by atmospheric oxygen in producing CO_2, which is then recycled in photosynthesis.

Green plants are the primary producers of the biosphere. They convert
solar energy into organic matter that is essential to all living organisms.
Forests, for example, cover about 10% of the earth's surface and fix nearly
one-half of the biosphere's total energy (5).

Figure 1.2 illustrates the net flow of energy in the biological cycle. The
energy that is fixed by the primary producers (i.e., plant life) is dissipated
in the form of heat in the respiration of plants, of plant consumers (i.e.,
the herbivores), and of successive echelons of carnivores and decay organisms. Virtually all nutrients are recycled, thereby permitting renewal
of the diverse plant and animal species that populate the biosphere.

The cycle shown in Fig. 1.2 is divided into two separate chains: the
grazing chain and the decay chain. On land, the decay chain starts with
dead organic material such as leaves, bark, and branches. In the water,
the chain originates from the remains of algae, fecal matter, and various
other organic debris. In some cases, the organic wastes are entirely

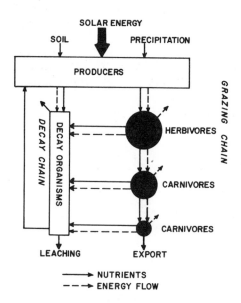

FIG. 1.2 The net flow of energy in the life cycle.

consumed by bacteria, fungi, and small animals, resulting in a release of carbon dioxide, water, and heat.

The decay food chain sometimes malfunctions. Under certain conditions all available oxygen is exhausted, resulting in incomplete decay. Products of these reactions are methane, alcohols, hydrogen sulfide, amines, and partially decomposed organics. This, of course, has an adverse impact on the grazing chain of the cycle.

The constituents of the earth's major ecosystems (i.e., the forests, oceans, marshes, estuaries, lakes, rivers, deserts, etc.) are influenced by a variety of parameters. The amount of living tissue that exists in any one of these ecosystems at a given time is determined by its net production, which in turn depends on the equilibrium between the processes of photosynthesis and respiration.

The major factors affecting photosynthesis are light intensity and duration, the availability of water and mineral nutrients, temperature, and concentration of available CO_2.

Respiration, the process by which CO_2 is released, is a continuous process, and as such is dominant in plants at nighttime, when photosynthesis is inoperative.

Carbon dioxide concentration in the atmosphere varies with the time of day. At sunrise, photosynthesis begins, marking a rapid decrease in the carbon dioxide concentration as plant tissues convert CO_2 into organic matter. Around noon, the ambient temperature rises and humidity decreases, causing the respiration rate to increase and the net consumption of CO_2 to decline. At sunset, photosynthesis terminates whereas respiration continues.

The net rate of fixation or net production of CO_2 varies with the type of vegetation. The amount of carbon from CO_2 consumed by phytoplankton in the oceans, on an annual basis, is roughly equal to the gross assimilation of CO_2 by land vegetation. The CO_2 that is consumed and the oxygen released are generally produced in the form of dissolved gas near the ocean surface. Most of the carbon in the sea is essentially self-contained; that is, the oxygen released is consumed by marine life which ultimately die and decompose, releasing CO_2 back into the sea. The amount of CO_2 dissolved in the upper layers of water in the ocean are in close equilibrium with the CO_2 concentration in the atmosphere at any point in time.

The CO_2 or carbon cycle on land is illustrated in Fig. 1.3. Carbon dioxide fixed by photosynthesis on land is recycled to the atmosphere by the decomposition of dead organic material.

The means by which carbon circulation is accomplished in the oceans is different from that on the land. Soil productivity is largely limited by the availability of fresh water and nutrients (primarily phosphorus). In the sea the primary limitation is the availability of inorganics. Phytoplankton requires plentiful quantities of phosphorus and nitrogen but also relies heavily on trace amounts of various metals (6).

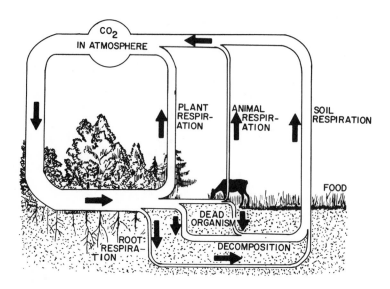

FIG. 1.3 The carbon cycle illustrates the processes of photosynthesis and respiration.

Phytoplankton is the primary fixer of carbon dioxide in the sea. Zooplankton and other tiny sea animals feed on the phytoplankton. These organisms in turn serve as the food supply for larger marine life. The majority of the ocean's biomass consists of microorganisms. The lifetime of these organisms ranges on the order of weeks to several months. This is in contrast to some plants on land with lifetimes measured in years. As these marine microorganisms complete their life cycle, they quickly decay and settle to deeper layers, becoming dissolved organic matter.

A portion of the settleable organics escape oxidation and merely sink to the lower ocean depths. This matter has a significant impact on the amount of chemical materials present as the deeper layers undergo circulation with surface layers at a slow rate. As the deep layers become enriched with nutrients and unoxidized organic matter, dissolved oxygen levels diminish. This is followed by an increase in dissolved CO_2, primarily in the form of carbonate and bicarbonate ions. The distribution of oxygen, carbon dioxide, and constituents has a direct bearing on the abundance of marine life and the availability of various chemical substances in the surface water layers. The rather slow vertical circulation in the ocean prevents the water from becoming saturated with oxygen. It also helps to enrich the deeper strata with bicarbonate and carbonate ions.

Figure 1.4 illustrates how carbon is circulated in the ocean and on land. Note that two distinct carbon cycles exist. In the ocean, the carbon cycle

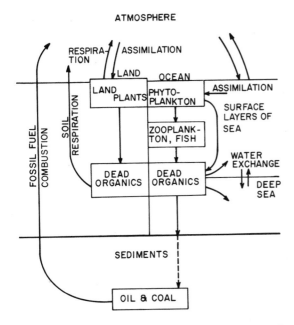

FIG. 1.4 Comparison of carbon cycles on land and in the ocean. There is, of course, interaction between the atmosphere and ocean.

is essentially self-contained as phytoplankton assimilates the dissolved CO_2 in seawater, releasing oxygen back into solution.

The foregoing discussion has not taken into account human impact on the various cycles in the biosphere. The incineration of large amounts of fossil fuels has greatly increased the amount of CO_2 in the atmosphere and oceans. Studies have shown that plants tend to grow faster when surrounded by an atmosphere enriched with CO_2 and, as such, man has inadvertently fertilized the environment by burning coal, oil, and natural gas (6). This has resulted in an increase in biomass on the land. This benefit has, however, been offset by man's tampering with the biological and geochemical balances of the biosphere.

BIOMASS AS A SOURCE OF ENERGY

On the land, major sources of biomass include unused standing forests, wood-bark logging residues, crop residues, wood-bark mill residues, manures from confined livestock operations, and agricultural residues accumulated at processing plants. It has been estimated that roughly 9.6

quads (Q) of biomass are generated annually within the first three sources
alonè and, of that amount, approximately 7 Q have the potential of being
viable energy sources. (One quad is approximately equivalent to 470,000
barrels of oil per day.)

There are certain constraints which currently limit the use of the first
three major sources of existing biomass. Supply constraints are one major
limitation, which includes the potentially higher value use of wood-bark
biomass for wood product manufacture. Similarly, the residual value of
crop residues to farmers is significant as a source of plant nutrients and
soil conditioners. The other set of constraints are related to problems of
potential users, e.g., public utilities, who are uncertain as to the reliability

FIG. 1.5 One form of land biomass. Here scientists are evaluating the
feasibility of double-cropping sweet sorghum (a fuel crop) and winter wheat
(a food crop) in the Midwestern corn belt. (Courtesy of Battelle Columbus
Laboratories.)

of long-term supply and the lack of demonstrated economic viability of
existing biomass sources (7, 8).

Biomass is essentially plant material, ranging in form from algae to
wood (refer to Fig. 1.5). Its energy content is relatively uniform, on the
order of 9000 Btu/lb. This represents a heating value roughly half to two-
thirds that of coal. This is actually a severe limitation in that shipping
over long distance makes it economically unattractive. The major advantages
of biomass as fuel are the following:

1. It contains negligible sulfur and, as such, does not create the air
   pollution problems that are associated with coal.
2. It generates little ash.
3. It is continually renewable.

The success of bioconversion systems will depend heavily on the economics associated with harvesting and biomass transportability, as well
as our ability to produce biomass continually without permanent damage
to the forests or depletion of the land.

As municipal, forest, or agricultural wastes, biomass is considered to
be a nuisance. Cities and towns in the United States are faced with enormous treatment and handling costs for waste removal and processing.
Incinerating these wastes creates environmental problems. As such, these
waste materials are available for bioconversion at relatively low cost.
Table 1.1 gives rough estimates of biomass wastes generated annually in
the United States (9, 10).

The primary energy sources that can be obtained from bioconversion
are illustrated in Fig. 1.6. A variety of processes are currently being
developed for the bioconversion of biomass into fuels. Figure 1.7 illus-

TABLE 1.1 Approximate Values of Waste Biomass Collected
in the United States

Sources	Millions Dry Tons/Year
Municipal	170
Raw sewage	60
Forestry	120
Field crops/processing residues	64
Manure	174
Total land biomass	588

Source: Data from Reed (9, 10).

BIOMASS
(forest, agricultural & municipal wastes)

GASEOUS FUELS	VOLATILE LIQUIDS	HEAVY LIQUIDS	SOLID FUELS
Wood Gas	Ethanol	Wood Oil	Charcoal
Synthetic Gas	Methanol		
	Gasoline		

FIG. 1.6 Primary sources of energy from bioconversion from the land.

strates several of the basic routes by which biomass can be used in fuel conversion.

Most naturally occurring biomass exists in diluted form. Drying or dewatering processes require considerable energy expenditure in order to transform biomass to concentrated forms.

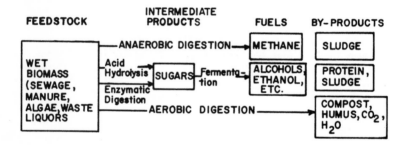

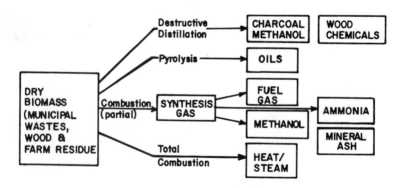

FIG. 1.7 Basic processes of biomass conversion to fuels.

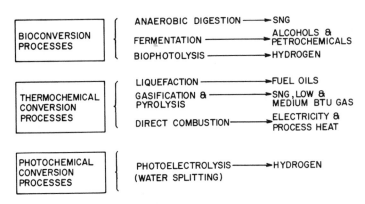

FIG. 1.8 Specific biomass conversion technologies currently being developed.

Some forms of biomass have been used as fuels for centuries. Figure 1.8 illustrates relatively new biomass conversion technologies that are currently under consideration. Agricultural residues are considered one of the chief sources of biomass for immediate and near-future energy production. Such residues consist of manures, straw, cornstalks, and various other farming by-products. Forestry residues include sawdust, bark generated in sawmills, and logging wastes such as branches and dead or diseased trees.

At present, assuming that biomass conversion technologies were fully developed, the annual production of land biomass could supply roughly 10% of total U.S. energy requirements. For biomass to play a more important role in the energy picture and have a certain impact on our rapidly growing energy demands, it will be necessary to supplement biomass production via terrestrial and/or aquatic production. Improvements in agriculture and silviculture and the use of presently uncultivated land as well as the nearby oceans can greatly increase biomass production.

The conversion of biomass to energy can be categorized thus:

1. Direct combustion of biomass for the production of process heat, steam, or electricity
2. The production of matter rich in carbon and/or hydrogen and poor in oxygen and nitrogen, the latter pair not contributing to high-energy content
3. Preparation of biomass as feedstock for work animals

Petroleum, natural gas, and coal have come to be our principal sources of fuel because of their high energy content. The high energy content associated with each of these is due to their chemical makeup, which is generally low in oxygen, nitrogen, and inorganic constituents. The major chemical

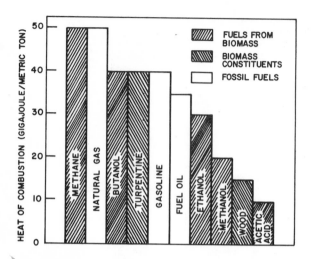

FIG. 1.9 A comparison of the heats of combustion of various biomass constituents, biomass fuels, and fossil fuels.

constituents in the conventional fuels are carbon and hydrogen, both of
which burn readily. The heat liberated from combustion is used to perform
work.

There are a large variety of plants that produce energy-rich chemicals.
Biosynthetic processes produce various by-products such as glucose that
may be rich in oxygen and carbon atoms. Carbohydrates, another product
of biosynthesis, are generally low in energy. Storage carbohydrates
[formula $(C_6H_{10}O_5)_n$] have slightly higher energy content because of the
loss of a water molecule from each glucose unit. Lignin, a sticky, gluelike
substance, has a considerably higher energy content than carbohydrates,
primarily attributed to its energy-rich aromatic structures (11, 12). Lipids,
terpenoids, and steroids are biosynthetically produced from acetate via
secondary metabolic processes (11). These substances are important as
food and pharmaceuticals; however, they do have potential as energy sources.
Figure 1.9 compares the heats of combustion of various biomass constituents, fuels from biomass, and fossil fuels.

Current practices in forestry and food crop farming are directed at the
economic production of only a specific portion of the plant. Fiber or lumber,
for example, is desired from wood. For food plants, growing is directed
solely for the edible portions. Genetic improvements, seed selection, plot
layout, and fertilizer application are all evaluated in terms of the economics
of the desired end products. On the other hand, production of biomass for
energy usually employs the entire plant, and the best species for total growth
are not necessarily those which are best for food, timber, or fiber processing.

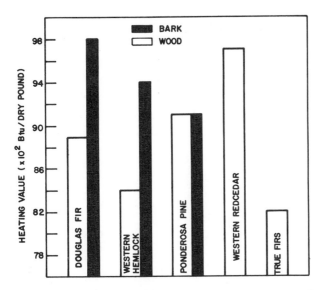

FIG. 1.10 Heating values of some typical wood residues from the Pacific Northwest.

The pulp and paper and logging industries are a good example of present applications of biomass conversion to energy. Current logging practices annually produce about 9.5 billion ft^3 of fiber in forests and primary and secondary manufacturing of wood plants generate 1.3 billion ft^3 of wastes per year (8,13). Over the past decade there has been a steady rise in the use of primary and secondary mill residues by the pulp and paper industry.

Other sources of wood residue include fiber-damaged and insect- or disease-ridden trees in urban regions, driftwood, construction wood wastes, and wood wastes from municipal solid wastes. This form of biomass is currently being utilized in a number of ways. For example, in the north-western region of the United States, hogged fuel is sold in bulk volumes, or in units. A unit of wood residue is that amount contained in a 200 ft^3 volume (14). One unit of hogged fuel contains about a dry ton of wood material, where the dry weight can vary from 2600 lb (hogged Douglas fir) to 1900 lb of sawdust. The availability of these wood residues is such that their utilization as a fuel supplement cannot be ignored. Figure 1.10 illustrates the heating values of some typical Pacific Coast wood residues.

The combustion of wood involves three processes:

1. Heat must be supplied to evaporate water in the wood fuel in order to effect combustion. The amount of water contained in dry wood

is an important parameter since as the moisture content increases, the heat content decreases proportionately.

2. Volatile hydrocarbon gases are evolved and mixed with oxygen, giving off heat.

3. More heat is released as combustion nears completion, with the oxygen reacting with the fixed carbon at high temperatures.

At first, these processes occur in succession; however, as heat is generated, the wood eventually sustains its own combustion and all processes occur simultaneously.

Many firms today use wood and bark residues as fuels. These residues are also burned for home heating in stoves, furnaces, and fireplaces. The heat of combustion of sander dust is used in veneer and wood particle drying processes and for the production of steam. The production of steam represents the largest industrial use of this type of biomass-energy conversion (14, 15). Steam is produced for heating, processing, and generation of electricity. Hog-fueled steam plants range in steam capacity from 10,000 lb/hr to an excess of 500,000 lb/hr (14).

One type of hogged wood fuel burning process widely used about 25 years ago employed a Dutch oven. In the first stage, water is evaporated from the wood and the fuel is gasified. During the second stage, furnace combustion goes to completion. This system is gravity-fed. As the hogged fuel enters the Dutch oven from above, it forms a conical pile. Today, more efficient and larger-capacity systems are utilized.

The fuel cell process is illustrated in Fig. 1.11. Wood enters the primary furnace compartment from overhead. Since the system is gravity-fed, the fuel drops into a water-cooled grate. The wood is first gasified in the primary stage, and gases are then passed into a secondary combustion compartment where complete combustion takes place. In the western regions of the United States, wood-fuel boilers of this type are widely used.

Another type of steam plant utilizing wood and bark biomass is the spread-stoker design. In this system a pneumatic or mechanical spreader feeds the wood fuel from above onto a grate in the furnace. As the fuel falls to the grate, it undergoes partial combustion while in suspension. Combustion is completed on the grate. This operation has been employed in small plants and has been successful in achieving steam rates ranging from 25,000 lb/hr to 500,000 lb/hr.

Other common designs include the inclined grate furnace system and a fluidized bed system. The former resembles municipal solid waste incinerator systems. Hogged fuel is introduced on the top section of the grate. The wood then passes through three zones of the grate. The first zone induces evaporation, and the second supports combustion. In the last section, combustion is completed and ash removed.

Fluidized-bed wood-waste heat recovery systems are capable of combusting hogged-wood residues with moisture contents as high as 55%. After

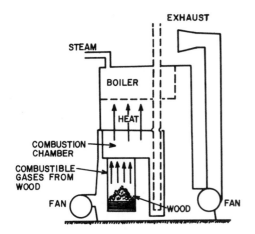

FIG. 1.11 The basic fuel cell process. Steam is generated by the combustion of residues from wood and bark, a commonly practiced bioconversion technique. (From Ref. 14.)

startup no supplemental fuel is necessary to maintain combustion. These systems are generally fully automated. Boiler efficiencies obtained from these units are comparable to conventional fuel equipment.

Most wood and bark fuels experience high heat losses because of the water evaporation required. In general, a fuel with 50% moisture content requires roughly 13% of the fuel's total heat output to evaporate the moisture (14). More than 25% of the final heat output is required for evaporation when the moisture content is 67%. As moisture content rises, flame temperature lowers and hence combustion is inhibited, resulting in a reduction of steam output of the boiler. Combustion cannot be self-sustaining if the moisture content is around 64 to 69%. High moisture content wood fuels usually must be dried prior to combustion or be burned along with a supplemental fuel such as oil and coal.

BIOMASS IN THE PAPER INDUSTRY

In general, the U.S. pulp and paper industry exhibits marked differences in the relationship between consumption of fossil fuels, electrical demands, and consumption of process wastes for fuel. Southern mills account for approximately 62% of the energy purchases of this industry. Process wastes, in particular the burning of black liquors for the generation of steam and electricity, provide roughly 50% of the fuel requirements. The North Central region purchases roughly 11% of the total energy used in the industry.

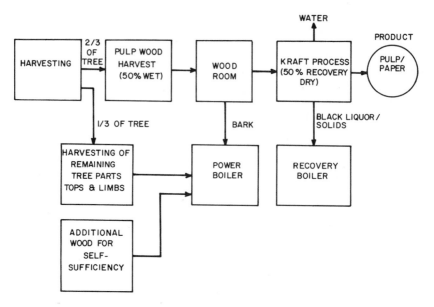

FIG. 1.12 Flow of wood in a pulp/paper mill.

Industries in this part of the country obtain on the average 15% of their fuel
requirements from process wastes (16).

Figure 1.12 illustrates the flow of wood material in a typical pulp/paper
mill. Roughly 75% of the pulp produced in this country is done by the kraft
process. Large steam boilers typically employed in the pulp/paper industry
are custom engineered for each specific application and assembled on the
site. Some components are shop fabricated, however. Large field-erected
boilers firing coal (primarily at North Central mills) have capacities in
excess of 10^6 lb/hr of steam, whereas industrial boilers firing wood alone
are limited to about half this capacity.

In general, field-erected boilers that are designed for wood firing require
larger combustion chambers which are significantly more expensive than
oil or gas boilers. Both wood- and coal-fired designs have relatively low
heat release rates.

The size and moisture content of residue fuels which can be incinerated
in field-erected boilers depend largely on the specific design of the combustion chamber, the air's preheat temperature, and the combustion properties of the fuel. As pointed out previously, moisture contents in wood fuels
in excess of 60% cannot sustain burning. In this range blackout conditions
can result. However, many systems can incinerate residues with moisture
levels as great as 55+%.

BIOMASS AS USE FOR FEEDSTOCK

There is great uncertainty over the future supply of petroleum feedstocks, particularly with regard to the impact of an inevitable shortage in the plastics and chemical process industries.

As will be discussed in subsequent chapters, a variety of chemicals can be derived from wood residues that are suitable as feedstocks. A large portion of the present production of silvichemicals relies heavily on oleo-resins, such as turpentine and rosin. Oleoresins are by-products of pulping operations. There is growing interest in the conversion of wood residues to specialty chemicals, particularly aroma products that are used by the flavor and fragrance industries. Wood can by hydrolyzed to its constituent sugars, which can be further converted to an array of fermentation or dehydration products. Lignin obtained as residue from hydrolytic processes can be converted to phenol by hydrogenolysis followed by hydrodealkylation. Figure 1.13 summarizes the principal chemicals that can be derived from wood and wood wastes. A brief description of the importance and use of some of these chemicals is given below:

Methanol

The principal use of methanol is a raw material in the production of formaldehyde, a constituent of synthetic resins used in the manufacture of plywood, chipboard, and plastics. Major resins produced are of urea-formaldehyde and phenol-formaldehyde compositions.

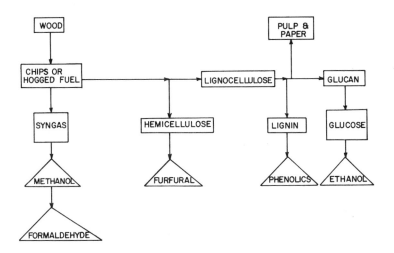

FIG. 1.13 Various chemicals that can be extracted from wood and wood residues.

At present, methanol is primarily manufactured from natural gas by the conversion of methane into syngas which contains hydrogen and carbon monoxide. Natural gas is composed of roughly 95% methane. Roughly 115 ft^3 of natural gas (or 4.9 lb) is required to produce 1 gal of methanol. The manufacture of methanol from wood residues is a rather complex system. A 10-day supply of wood waste requires approximately 10 acres of storage area and extensive mechanical equipment for handling. Crude gas from reactor-gasifiers requires a series of steps to clean the gas, remove hydrocarbons and nitrogen, and convert a portion of the carbon monoxide to obtain the proper ratio of hydrogen to CO for synthesis (17). The reforming of methane into syngas is an endothermic reaction and as such has an associated energy input. The reaction requires about 15% of the total feed energy input, and roughly 15% additional energy is necessary to drive the multiple-stage compressor trains. A second alternative is the conversion of coal to syngas. This latter approach has a greater efficiency than wood waste, as coal has a higher carbon content and a lower oxygen content than wood.

Ethanol

Ethanol is a chemical that finds wide use as a starting feedstock for the synthesis of drugs and medicine. It is also widely used as a raw material for production of ethers, glycols, and ethylamines. At present, the primary method used in the production of synthetic ethanol is the direct hydration of ethylene and water. This reaction is carried out at conditions of about 1000 psi and 500° F in the presence of a phosphoric acid catalyst. Ethylene is a basic building block in the petrochemical industry, where it is produced by the cracking of ethane and propane, or of naphtha, and petroleum fractions.

Sugars

The conversion of wood or wood residues into sugar involves hydrolysis in the presence of a dilute mineral acid catalyst at temperatures and pressures as high as 385° F and 225 psig, respectively. After the recovery of the sugars, six-carbon sugars can be fermented by yeast into ethanol. A second mineral acid hydrolysis step can convert the remaining five-carbon sugars into furfural (discussed in the next subsection). At present, the use of wood wastes in the manufacture of sugars and ethanol still requires a rather high capital investment.

Furfural

Furfural is hydrogenated to furfuryl alcohol, which is added to urea-formaldehyde resins for foundry core binders and special adhesives; it is also employed in the extraction of butadiene and wood rosin refining. Fur-

fural is a five-carbon heterocyclic organic compound synthesized commercially from C_5 carbohydrates (e.g., pentose sugars from oat hulls and corn
cobs). At present there is no known practical synthesis scheme from wood.
A furfural plant based on wood wastes requires roughly 33% of the residue
for plant energy demands. The remaining biomass represents a potential
source of various other chemicals or a substantial energy source.

Phenol; Benzene

Phenol finds its main use in the manufacture of phenolic resins via combination with formaldehyde. In the past it was primarily obtained as a by-
product of the coal industry. Presently, virtually all phenol is manufactured
from petrochemical feedstock.

Sugars from hemicelluloses and various chemicals from lignins appear
to be highly promising in the paper industry, although pulping processes
may have to undergo modification to make sugars usable (18). Considerable
research into the various usable chemical derivatives of lignin is currently
underway with promise in the production of benzene and phenol.

Other Research

Biomass also has great potential as an auxiliary feedstock in a coal-based
syngas plant. As noted earlier, one of the most important chemicals in the
U.S. industrial economy is ethylene. Ethylene serves as a building block
for a multitude of feedstock chemicals, and therefore considerable investigation is being done with the aim of producing ethylene from biomass.

Another type of biomass utilization study is in the area of conversion of
natural oils to feedstock. A great deal of research is being conducted by
the U.S. Department of Agriculture (USDA) into preparing plasticizers from
cottonseed oil derivatives (19). These derivatives might replace dioctyl
phthalate in polyvinyl chloride (PVC). The plasticizer under consideration
by the USDA at the Southern Regional Research Center is based on a fatty
amide derived from the oleic portion of the oil. All but one double bond of
the amide is eliminated by selective hydrogenation. This new site is made
available for reaction of the molecule with PVC. It is worth noting that the
oleic molecule is present in a variety of natural oils, and so the process
and economic considerations are not solely dependent on cotton (19).

Tallow (inedible animal fat) is another potential source of feedstock,
although strictly speaking it is not considered to be biomass. It has the
potential of supplying chemical derivatives useful in plastics and surfactants.

Research efforts by the USDA are also being directed toward cereal
grains. Starch, for example, can be used as a filler and reinforcing agent
in polyester-based urethanes, and starch-derived polyol glycosides can be
employed in making rigid polyurethane foams in the plastics industry.

Some success has been achieved in making nylon from vegetable oils.
Nylon-9, for example, has been prepared from soybean oil using a four-step

process. In France, a process for the manufacture of nylon from natural oils is already on a commercial basis. The process, which uses castor oil, is called Rilsan. Castor oil is a mixture of triglycerides, principally ricinoleic acid. Alcoholysis produces methyl ricinoleate, which then undergoes a cracking operation to form methyl undecylenate. Hydrolysis and amination reactions follows, producing the monomer, which is 11-aminoundecanoic acid (19). The polymer, nylon-11 is obtained by a polycondensation reaction. Upon further processing, a fine powder suitable for molding, extrusion, or coatings formulations is produced.

    Biomass as used for feedstock will require varying degrees of pretreatment. The amount and complexity of preprocessing necessary will be determined by both technical and economic constraints and will vary depending on the types of waste and natural materials used, as well as the type of process reactors required. Size reduction of wastes may be a necessary pretreatment step in order to ensure troublefree reactor-feeder operation. Specifically, it may be necessary to prevent bridging from occurring in certain types of reactors and to increase the heat-transfer efficiency of a reactor. Size reduction may also be necessary for recovery of metals and in separating organic fractions from municipal refuse. Reactor throughput may be increased by cubing or pelletizing wastes (called densification), such as cotton-gin trash or rice straw (20). Drying or dewatering operations may also be necessary in preparing low-moisture content feedstocks.

FOODSTUFFS AND BIOMASS

The beginnings of agriculture occurred some 10,000 years ago with the domestication of plants and animals. Prior to this, humans were herbivores, then scavengers, and then predators. Through the ages man has essentially shaped the biosphere to meet his own needs, by altering the earth's plant and animal populations. Today, farmlands cover some three billion acres, or roughly 10% of the earth's total land surface. Roughly two-thirds of the cultivated cropland is devoted to cereals. For wheat alone, nearly a million square miles is used.

    Man's agricultural endeavors and animal husbandry have not only altered the relative abundance of plant and animal species but their global distribution as well. The interchange of crops and animals between different parts of the world has increased the earth's capacity to sustain human life. Through genetics man has been able to improve on the natural evolution of plants and animals via selective breeding. This has made possible the development of various cereals and plant species more tolerant to cold, more resistant to disease and drought, and higher in yield.

    The controlled biomass generated is consumed by man either directly or indirectly by first feeding crops to domestic animals. Some of the food for domestic animals comes from pastures, of course, and not from con-

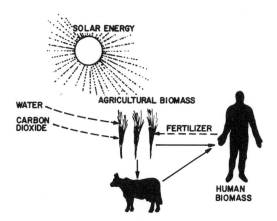

FIG. 1.14 Biomass as consumed by animals and humans. In man's case, biomass can be consumed directly or indirectly by first feeding it to livestock.

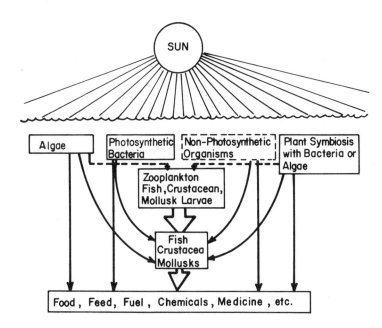

FIG. 1.15 Basic utilization scheme in marine systems.

trolled croplands. Man also obtains food from sources other than agriculture, for example, fishing. The result is that man's life cycle involves a human biomass chain. This is illustrated in Fig. 1.14.

As indicated earlier, the driving force behind the life cycle is solar energy as harnessed by photosynthesis. The supply of solar energy—both that portion stored in fossil fuels and that taken up daily and converted into food energy by plants—enables an advanced country to be fed with only about 5% of the population directly employed in agriculture (21).

Man's harvesting of the vast amounts of biomass in the sea is a relatively recent development. About 71% of the earth's surface is covered by ocean. Figure 1.15 illustrates the basic scheme by which man is beginning to obtain foodstuffs, fuel, chemicals, and other products from marine systems.

The marine environment offers a number of unique characteristics when considering food production. Organisms have a proportionately larger surface area for the absorption of nutrients. The temperature regime is much less prone to strong perturbations, in particular freezing. The flow of water created by tides or currents provides for nutrient input and recycling of metabolic products and, in addition, provides a three-dimensional space for growth of algae; these organisms are in turn fed upon by higher orders of marine life.

BIOMASS FROM THE SEA

Marine photosynthetic resource culture has had a relatively short history in comparison to land agriculture. Most of the technologies presently available emanate from Asian nations.

Three classes of macroalgae are cultured throughout the world; these are the brown algae (Phaeophyceae), the red algae (Rhodophyceae), and green algae (Chlorophyceae).

Today in Japan, for example, prefectural and municipal laboratories culture species of red algae in tanks. Marine farmers dip their culture nets into these tanks, which allows seeding directly onto the netting (22, 23).

Brown algae are harvested on the largest scale, worldwide. In this regard, the important genera of brown algae include the giant kelp Macrocystis, as well as Undaria and Laminaria.

Green algae make up only a small portion of the total algae harvested in the world. The principal countries that harvest green algae are Korea, Japan, Mexico, and Argentina.

The amount of algae used as a food source for fish and shellfish under culture is estimated to exceed about 600 million tons per year (24). There are a variety of organisms, particularly mollusks, that are efficient at straining phytoplankton and converting it to higher-quality protein (22).

Microalgae are consumed in many regions of the world. They are used in a variety of forms, including as salad vegetables, as cooking vegetables, and in soup (25, 26).

TABLE 1.2 World Harvest of Marine Plants (Metric Tons)

Year	Brown Algae	Red Algae	Green Algae	Other Plants	Total
1971	483,600	375,500	1,100	80,300	940,500
1972	506,500	322,800	700	89,800	919,800
1973	586,900	450,300	900	95,400	1,133,500
1974	696,567	525,654	2,237	94,834	1,319,292
1975	633,182	422,424	2,487	104,857	1,162,950

Source: From Ref. 27.

Although there appears to be considerable nutritional value that can be derived from marine biomass, the use of marine algae and plants as food-stuffs for humans has gained only limited acceptance worldwide. Algal and marine plant products have regionally limited use as food in Asia, Northern Europe, the Pacific Islands, and some regions of South America. Table 1.2 lists the world's harvest of marine plants in metric tons over a five-year period.

Marine plant life has been found to contain a multitude of chemicals and proteins. Various species of seaweed contain anywhere from 5 to 50% crude protein on a dry weight basis (28).

The protein content found in photosynthetic microorganisms, which include unicellular algae and photosynthetic bacteria, has been reported in the range of 10-65% on a dry weight basis (22).

The protein content of marine seagrasses has been determined to be in the range of 10 to 20% of dry weight. Various algal species have been found to have carbohydrate content in the range of 10-65% of dry weight (22,29).

Photosynthetic organisms have been reported to show the capability to produce hydrogen gas (30-32). The primary drawback appears to be that the solar energy conversion efficiency of the process is relatively low. There are, however, a number of advantages to this approach to hydrogen photoproduction in comparison with thermochemical and physical methods discussed later. These advantages include:

1. Such a biological system may be operated at relatively low temperatures (in the range of 10-40°C). Chemical or physical production of hydrogen is operated at high temperatures (i.e., 400-1000K).
2. The primary feed to such a system would be solar energy and a hydrogen donor such as water.
3. This is a clean source of fuel. The process does not involve the generation of pollutants, as in the case of fossil fuel refineries.

Marine plants have also shown great promise as a source of methane. Capturing and storing solar energy via the photosynthetic process is highly attractive because of the virtually inexhaustible supply of solar energy and the fact that this stored state is renewable. As indicated earlier, however, the low incident flux of solar energy over the earth's surface may be a limiting factor. Estimates of the incident flux show the order of magnitude to be about 500 cal/cm^2 per day (33). An additional complication is the upper limit of capture of solar radiation by plants, estimated to be about 8% (33).

Life cycles of marine algae vary considerably among species. They display developmental patterns which are significantly more complex than those of land plants. At present there are two general approaches being considered in the conversion of marine plant life to methane, namely, anaerobic digestion and hydrogasification. The former can almost be considered established technology. Investigation into hydrogasification of biomass is still in its infant stages. Both technologies are discussed in subsequent chapters.

The U.S. Department of Energy [DOE—incorporating the Energy Research and Development Administration (ERDA)] is supporting a variety of research projects directed at the production of energy sources from seaweed. One area under investigation involves a combined sewage treatment/aquaculture process in which marine plants are grown on wastes, producing methane and various other hydrocarbons. Another area of research has the potential for producing large amounts of methane from huge volumes of giant kelp that are implanted on man-made structures 500 ft or more deep, off the coast of southern California (34).

Literally thousands of species of seaweed are found along most of the coastlines throughout the world. At present, less than 100 species are utilized. Seaweeds, or marine algae, range in size from 400-ft strands (giant kelp) to single-celled organisms.

Seaweeds have found wide use as animal fodder and crop fertilizers, and and limited use as food for humans.

Principal products of seaweed processing are agar, algin, and carrageenan. Agar [more commonly called agar-agar (the original Malay name)] is processed from red seaweeds. It comprises a class of vegetable gums (called polysaccharides) that are employed in suspending, thickening, and stabilizing solutions. Agar has the properties of high gel strength at relatively low concentrations, exhibits low viscosity in solutions, is resistant to heat degradation, and is transparent. The bakery and confectionary industries also use agar widely.

Algin is a vegetable gum. It is one of the chief constituents in many species of brown algae (kelp). It has gelling, thickening, suspending, emulsifying, and water-holding properties. It is primarily employed in the textile industry and also in the pulp and paper, pharmaceutical, and food-processing industries.

Carrageenan is obtained from red seaweeds and is used as a thickener, stabilizer, and gelling agent. It is employed by the food industry. Other applications include texturing binders in toothpaste and in gel bases to control the release of fragrance in solid household air-fresheners.

At present there is worldwide interest in finding new chemical derivatives from marine plants, particularly those chemicals suitable as feedstock.

Nuisance Growths

A considerable amount of unused biomass populates freshwater bodies throughout the world. These are the so-called aquatic weeds and are considered to be nuisances. Aquatic weeds can affect water use adversely by blocking canals or pumps in irrigation projects and interfering with hydroelectricity production and boat traffic; they can also cause flooding by clogging rivers and canals. These types of problems are particularly acute in tropical countries where warm water and increasing numbers of dams and irrigation projects are ideal for such aquatic plant growth. Problems associated with nuisance growths have worsened because of waters becoming enriched by fertilizer runoff and from an overabundance of nutrients derived from human and agricultural wastes.

Various freshwater aquatic weeds can be responsible for spreading waterborne diseases. For example, they disperse water snails that are responsible for schistosomiasis, a debilitating disease prevalent in many underdeveloped countries. Aquatic weeds also foster malaria, encephalitis, and other mosquito-borne diseases (35).

Nuisance growths do, however, constitute a vast source of biomass, having both potential energy and feedstock value. They are essentially a highly productive crop that requires no tillage, fertilizer, seed, or cultivation. Specifically, they have potential uses as animal feed, human food, soil additives, fuel sources and wastewater treatment.

Considerable research has been directed toward the water hyacinth. The National Aeronautics and Space Administration (NASA) is working on converting water hyacinth and other aquatic weeds into a biogas rich in methane. Methane is the main ingredient in natural gas, which is used worldwide as fuel. The recovery of fuel from aquatic weeds has interesting implications, especially for rural areas in developing countries. As many developing nations have an apparently inexhaustible supply of aquatic weeds within their borders, this potential energy source could be highly important.

Aquatic weeds are converted to biogas by capitalizing on decay by anaerobic bacteria. Methane-producing bacteria are common in nature (for instance, in the stagnant bottom mud of swamps, where they produce bubbles of methane known as "marsh gas"). If bacteria are cultured on water hyacinth in a tank that is sealed to keep out all air, they produce a biogas composed of about 70% methane and 30% carbon dioxide (35-37).

The high moisture content of aquatic weeds is an asset for a fermentation operation. Hence, this is one method of aquatic weed utilization that does not require dewatering, which is economically advantageous.

Based on NASA's findings it appears that the water hyacinth harvested from 1 hectare will produce more than 70,000 m^3 of biogas. Each kilogram of water hyacinth (dry-weight basis) yields about 370 liters of biogas with an average methane content of 69% and a calorific (heating) value, when used as a fuel, of about 22,000 kJ/m^3 (580 Btu/ft^3). In contrast, pure methane has a calorific value of 33,300 kJ/m^3 (895 Btu/ft^3); see Refs. 3 and 5.

Biogas burns readily and can be employed in almost all the same applications that natural gas is used for. Before the gas can be compressed into cylinders for use on equipment, carbon dioxide must be removed. In addition, although anaerobic digestion has been used for about 70 years to transform sludge in sewage treatment plants, it is attracting increased attention and research activity today. Several thousand digesters (air-tight tanks that permit the growth of anaerobic bacteria) are in operation around the world. These units, however, are directed at processing animal manure or human wastes mixed with vegetable waste. It remains to be determined whether, without any animal manure, aquatic weeds can be digested to produce substantial amounts of methane.

Methane-producing bacteria must be nurtured as they require such nutrients as nitrogen, potassium, and phosphorus. The work at NASA has shown that water hyacinths provide these elements in the quantities and proportions adequate for good growth of the bacteria and for good gas production.

The production of biogas removes carbon from the ferment, but there is little loss of the other elements. These remain as a liquid sludge that is itself an organic fertilizer and soil conditioner equivalent to compost, a useful by-product. Weeds usually have to be chopped or crushed to make them more available for bacterial attack in the fermentation process. Although simple, inexpensive equipment can be used for fermentation chambers, they must be carefully constructed. Methane-producing bacteria cannot survive if oxygen is present in the culture medium. A disadvantage is that the air that is introduced when the chamber is loaded with weeds results in a lag before anaerobic digestion can begin. During this lag, which can be up to 10 days, the oxygen is used up by aerobic bacteria, which produce carbon dioxide and not methane. With the oxygen gone, anaerobic bacteria take over, but biogas always contains a substantial amount of carbon dioxide, a nonflammable gas that dilutes the heating value of the methane (35).

Biogas production is not fast, with time on the order of 10-60 days. Furthermore, the initial start-up is slow because the bacteria take time to build up a population of sufficient size to ferment the weeds. Experience with existing biogas generators (using a medium of animal, human, and vegetable wastes) has shown that the bacteria are sensitive and their digestive processes easily disturbed.

Factors that must be taken into account in methane production from water weeds include:

1. Temperature Large temperature fluctuations can harm the bacteria; consequently, temperature must be maintained relatively constant.
2. Nitrogen Sufficient quantities of nitrogen must be present in the feedstock to support bacterial growth.
3. Mixing The fermentation reaction requires periodic agitation or mixing.
4. Acidity Various acids generally accumulate during digestion. Acidity can suppress bacterial growth; hence, addition of alkali is required for neutralization.

The conversion of aquatic weeds to biogas has only been attempted in a few laboratories. It is new and largely untested under conditions normally encountered in developing nations; however, the results are potentially so important that there is a need to test the system in weed-infested developing countries worldwide.

Besides water hyacinth, plants such as duckweed, water milfoil, hydrilla, alligator weed, and various algae have shown initial promise as feedstocks for methane-producing bacteria (35). Much research is needed to reduce the fermentation time and to optimize yields. These problems are discussed in depth in later chapters.

2

Energy and the Biosphere

THE EARTH'S ENERGY CYCLE AND RESOURCES

There is a constant flow of energy into and out of the earth's surface environment. The principal source of energy is solar radiation. Small amounts of thermal energy from the earth's core and tidal energy from gravitational forces of the earth, moon, and sun supplement this energy form.

At the surface of the earth, a relatively small amount of matter is accounted for in living organisms such as plants and animals. As described earlier, plant tissue captures a small proportion of the incident solar radiation and stores it chemically via photosynthesis. This biologically stored energy is released by oxidation at a rate that matches the rate of storage. Under conditions of incomplete oxidation and decomposition, some of this organic matter has been transformed into our fossil fuels.

The inward flow of energy through the earth's surface environment is illustrated in Fig. 2.1. For solar radiation, the influx is normally expressed in terms of the rate of energy flow across a unit of area that is perpendicular to the radiation and outside the earth's atmosphere at the mean distance between the earth and sun. The total solar energy intercepted by the earth is on the order of 1.39 kW/m^2 (38).

Heat conduction from the earth's core, estimated from geothermal gradients and thermal conductivity properties of rock layers, has been determined to be on the order of 0.063 W/m^2. Tidal energy has been estimated to be a small power source compared to other available forms of energy. Roughly 99.98% of the total power influx into the earth's surface environment is accounted for in solar radiation, or [as noted by Hubbert (38)] solar radiation represents roughly 5000 times the energy input from all other sources combined.

Of the solar radiation intercepted by the earth, roughly 30% is reflected directly back into space, 47% is absorbed by the biosphere, taking the form

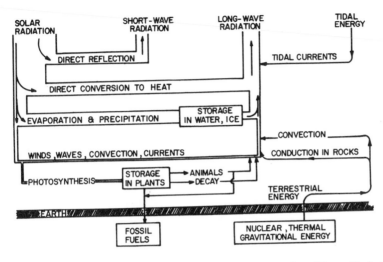

FIG. 2.1 The flow of energy to and from the earth. (From Ref. 38.)

of heat at ambient surface temperatures, and 23% is utilized in the processes
of evaporation, convection, precipitation, and surface runoff of water in
the hydrologic cycle (illustrated in Fig. 2.2). A small portion of solar
energy is utilized in driving atmospheric and oceanic convections and circu-
lations. As noted in Chap. 1, another fraction, even smaller, is captured
by the chlorophyll of plant leaves.

Fossil fuels began forming about 600 million years ago. Their formation
is still continuing but probably at no greater rate. The rate at which these
energy sources have been tapped by man have approached an exponential
growth rate in this century alone. For industrial uses the world's consumption of energy is doubling every 10 years (38,39).

Today's industrial society is entirely dependent on high rates of consumption of natural gas, petroleum, and coal. The large rise in consumption
rate, particularly in the United States, is attributed to household, commercial, and transportation uses as well as industrial requirements. In general,
all energy conversion systems are inefficient. In electricity generation,
losses are experienced at the power plant in transmission as well as at the
point of application. Inefficiencies in our energy conversion schemes result
in a tremendous loss of energy each year. As an example, the waste heat
from U.S. power generation is estimated to be of sufficient quantity to heat
every home in this country (40).

World energy consumption cannot increase indefinitely. The world use
of energy reached 2×10^{17} Btu/year in 1970 (41), and the average per capita
use of energy throughout the world increased 57% from 1950 to 1970, reaching 53 million Btu per capita per year in 1970 (41). The United States alone

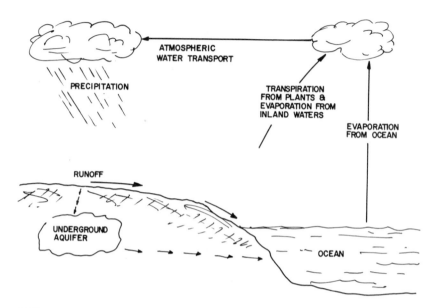

FIG. 2.2 The hydrologic cycle.

used 35% of the total or about 68×10^{15} Btu/year at an average of roughly 340 million Btu per capita per year, and the other industrialized nations using another 35% at an average of about 150 million Btu per capita per year (40). The developing countries of the world used approximately 55×10^{15} Btu/year, at an average of perhaps 20 million Btu per capita per year in 1970, but the greatest increase in per capita use occurred in these areas: 73% increase in Africa, 122% increase in Latin America, and 197% increase in Asia between 1950 and 1970 (40).

Wilcox (42) has noted that if such increases were to continue escalating at such rates for the next two centuries, by the year 2170 human activities on this planet would be generating heat at a rate equal to 10% of the solar energy input and the polar ice caps would probably begin melting; moreover, by the year 2230 the oceans would be close to boiling. But Wilcox also notes that it is possible for the per capita demand to become saturated and that total energy consumption will conceivably level off in the next 100 to 125 years. Kahn (43) has suggested that a "postindustrial" society is developing in which most human occupations will consist of services, recreation, and other "quarternary" activities. Energy use per capita thus might be no greater or even less in such a society than it is now for upper-middle-class persons in the United States; however, Kahn notes that at the 5% economic growth rate attained by the world during the 1960s, almost every country should be as well off in 100 years, in terms of gross national product

per capita, as the United States is today. As such, in about a century the
world could be approaching an era of stable population, with a relatively
flat income and energy use distribution, and enough affluence so as not to
want to increase either.

In this country, signs of energy saturation are already apparent. Between 1960 and 1970 the average per capita use of energy by families in the
upper quarter of income distribution increased by only 2% as compared with
3% for the national average (44). During World Wars I and II and the Korean
and Vietnam conflicts, substantial increases in energy use occurred. Zraket
(45) has suggested that a 2% annual increase in total energy use in the United
States is achievable within 20 to 30 years and that, in the long run, major
reductions in energy use can be attained by more efficient energy generation
and end-use systems without limiting economic growth.

Currently, major inequities exist in the United States in terms of both
income and energy consumption. As such, it is possible to envision an
increase by the poorer segments of the economy without a proportionate
increase in the overall energy use. Thus, a value of 500 million Btu per
capita per year can be regarded as a value representing a very healthy
economy, even for the year 2100. However, recognizing the potential desire
of everyone in most developed and developing nations to achieve and use the
goods and services available to the upper middle class, a figure twice as
large as that (a rate roughly three times greater than the present U.S. level)
can be projected as the maximum energy required to fuel a very affluent
world society. Thus if, for example, China, with its vast population and
present relatively low per capita use of energy, were to become an industrialized energy consumer on a Western model, the energy use in the world
would escalate at an even more rapid rate than already feared (46).

Large-scale dependence on fossil fuels as a primary source of energy
began around the latter part of the nineteenth century. Previously, solar
radiation, wind, and biomass were the basic energy sources used throughout human history. Today, new technologies are being formulated based
on sources other than fossil fuels. It is worthwhile reviewing the major
technologies and energy sources available to modern man before discussing
the potentials of biomass.

Nuclear Fission

At today's prices, there is sufficient uranium to fuel the nuclear reactors
projected to be constructed in the next decade for only about 30 years (47);
at prices five to ten times higher than at present, perhaps ten times as
much uranium would be available (48). Teller (49) believes that there is
adequate uranium for the foreseeable future even utilizing non-breeder-type
reactors. If the breeder reactor is developed in the near future, roughly
100 times more uranium would be available and possibly the very large
world resources of thorium could be "burned." In this case, there is no

question that sufficient nuclear fuels exist to supply the world for many
centuries.

A major question remains unanswered, however: Can radioactive wastes
be collected safely and stored to prevent man-made radiation emitted to the
biosphere from reaching levels that are in excess of the natural radiation
background? The environmental impact of nuclear fission is probably its
most serious drawback as a possible fossil fuel alternative.

Nuclear Fusion

The feasibility of production of useful energy from a nuclear fusion system has yet to be demonstrated on a practical level. It is conceivable that
a significant portion of man's energy needs might be met by the fusion process sometime in the twenty-first century. Theoretically, there is sufficient
fuel in the oceans and earth's crust to furnish energy for fusion at the rate
of 2×10^{19} Btu/year for a virtually indefinite period (48). The potential
release of radiation from projected fusion power plants is estimated to be
on the order of 10,000 times less than for a nuclear fission reactor (46).
In addition, there are no long-lived fission products to dispose of from such
a process.

Geothermal Energy

There is a continuous flow of heat from the interior of the earth to the
surface at an average of about 0.02 Btu/(hr $\cdot$ ft^2), or about 0.06 W$_{thermal}$/m^2
(46). Gustavson estimates that of this total useful heat flow, only about
10^{11} W (thermal) (48) could be utilized worldwide. Cheremisinoff and
Morresi (2) note other estimates. An energy system that would utilize all
the heat flowing out from 1 square mile would have a continuous supply of
140 kW$_{thermal}$ (roughly 40 kW$_{electrical}$, assuming a conversion efficiency
of 30%).

Geothermal systems are currently in use in Italy, Japan, Iceland, New
Zealand, and the United States, as well as several other countries. Approximately 300 MW of electrical generating capacity exist in the United States
today. Many thousands of acres of likely geothermal sources under public
lands were recently leased to public utilities by the U.S. Department of
the Interior. The National Science Foundation has obtained an estimate
that at least 10^5 MW of generating capacity could be developed in the United
States by the year 2000. There appears to be no doubt that geothermal
energy will be an important source of heat and electricity in many countries
in the future. However, it is unlikely to be the major source of energy for
the world primarily because of the limited availability of near-surface hot
rock or dry steam sites. In addition there are potential pollution problems
which include release of hydrogen sulfide and other noxious gases as well
as salts.

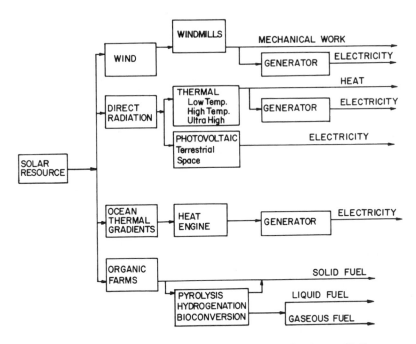

FIG. 2.3 Scheme for prospective development of solar radiation.

Solar Energy

During preindustrial times and early stages of industrial society, solar radiation supplied a major portion of man's total energy needs. Figure 2.3 illustrates four potential sources of solar energy: (a) direct radiation; (b) wind; (c) biomass/bioconversion; and (d) ocean thermal gradients.

The sun radiates a relatively narrow band of wavelengths (between 0.22 and 3.3 μm). At the outer fringes of the earth's atmosphere, solar radiation falls on a surface perpendicular to the sun's rays with an intensity of 442.2 Btu/(hr · ft^2). This quantity is referred to as the solar constant.

At any point on earth, the intensity and amount of solar radiation received at the surface varies with the season, latitude, and atmospheric transparency. (Refer to Fig. 2.4 for a distribution of solar energy in the United States.) Because of the variability of solar radiation, energy storage and/or backup power will be necessary to provide a failure-free capacity during nighttime or when the sun is obscured. Another drawback is that, because of the low density of solar radiation, large surface areas are necessary for energy collection.

Direct usage of solar radiation has been suggested for two approaches: (a) direct heating (e.g., as in a home water heater); and (b) heating a work-

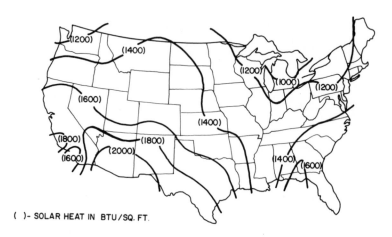

( )- SOLAR HEAT IN BTU/SQ. FT.

FIG. 2.4 Distribution of solar energy in the United States.

ing fluid that may be utilized in developing power in a heat engine or in transferring heat to an ultimate receiver. The latter approach is known as thermal conversion and can be divided into three categories:

1. Low-temperature radiation
2. High-temperature concentrators
3. Ultrahigh-temperature concentrators

Refer to Cheremisinoff and Regino (50) for a detailed treatment of this subject.

It has been estimated that some six to ten times the amount of energy necessary to heat a building radiates down on its roof each year. If solar energy is employed in heating a working fluid, a significant fraction of the energy demands of the house can be satisfied by various conversion processes. The use of solar energy for space heating has now been employed in edifices for several years; however, long-term and continuous experience with this approach is limited. The approach involves the use of low-temperature collectors. One application of a low-temperature solar collector is illustrated in Fig. 2.5. A working fluid, water, is used to supply space and water heating directly. The hot water can be employed to operate thermally driven refrigeration systems as well as to store heat for nighttime use. Fluid temperatures normally operate in the range of 200-250° F.

In high-temperature concentrators, solar radiation is concentrated with reflecting surfaces. These reflecting surfaces can be of the form of parabolic troughs or two-dimensional parabolic mirrors, with water pipes running along their focal lines. Steam temperatures as high as 600° F can be attained with these systems. This approach can be used for space heating, absorption refrigeration, or to produce steam for industrial processes.

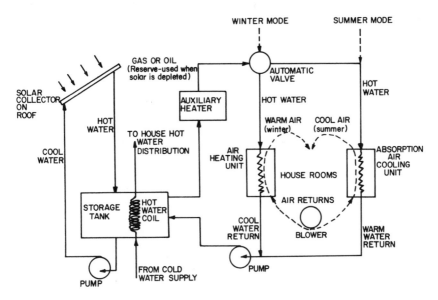

FIG. 2.5 Use of a low-temperature solar collector in residential heating and cooling.

Solar-generated steam has been suggested for use in large-scale electrical generation. According to this scheme, the output of a large number of collectors would be combined to operate conventional turbine generators. Again, our ability to predict the amount of solar radiation that will be available nationwide at any given time is limited; this uncertainty naturally has a bearing on the reliability of proposed central solar power plants.

Precisely contoured parabolic reflectors have been suggested for ultrahigh-temperature solar concentrators. By such means, temperatures as high as 5000° F may be attained (51). These systems have the advantage of relatively high conversion efficiencies in steam engines and turbines. Reflectors differ from trough-type concentrators in that they are true paraboloids which can reflect all incoming solar energy to a single point.

Other direct uses of solar energy include photovoltaic cells. These consist of a silicon solar cell which converts solar radiation directly into electrical energy. Today solar cells are the major source of power for space vehicles. At present, because of high cost, they have rather limited use for terrestrial power generation. To be commercially practical for large-scale power generation, solar cell costs must be reduced by a factor of roughly 1000 (51). It is generally agreed that the technical advances necessary for satellite power stations to become feasible for electricity generation are so formidable that such systems are not likely to play a

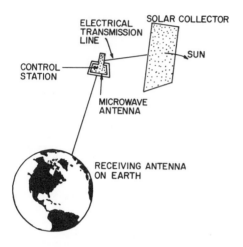

FIG. 2.6 Conceptual model of a satellite solar power station for electricity generation.

significant part in supplying energy demands in the near future. (Figure 2.6 illustrates conceptual design of a satellite solar power station.)

The conversion efficiency of solar-based systems impacts on the size of facilities required to generate a specified amount of energy. This, in turn, impacts on the capital investment and overhead costs.

For direct heating applications with either flat plate collectors or parabolic systems, conversion efficiencies reach a maximum somewhere between 60% and 70%, depending on the specific application. For electricity generation, the combined efficiency of collectors, storage systems, heat engines, and various equipment necessary for electrical power generation is not likely to exceed 20% (51). For photovoltaic generators, efficiencies are not expected to exceed 15-20% with present technology.

The great advantage of using solar energy is that no additional heat or residue (pollution) is released at the earth's surface (unless a collector in space is employed). Disadvantages of using solar energy include the tremendous areas involved in collecting it and its diurnal nature, which imposes the need for some form of energy storage. In general, the net heat residuals for solar electrical power generation are negligible; however, it should be noted that there is the potential of localized thermal pollution.

The operating costs of solar power plants are expected to be low, but initial capital investment will represent a major portion of generating costs. This makes solar-generated energy several times more expensive than fossil or conventional nuclear power at present fuel costs. Installation costs are, however, expected to be comparable with those of breeder nuclear power plants.

Ocean-based systems for producing electricity and fuels may eventually be feasible. The current state of the art is the ocean thermal difference technique with an efficiency of only 1-3% in converting thermal differences in the ocean into electricity. Future devices may be based on photovoltaic or photogalvanic effects, perhaps with higher efficiency. If 1% is taken as the overall efficiency of converting sunlight to electricity, roughly 7 million square miles of ocean (about 5% of the total ocean area) would be serving as a collector to produce one 10^{18} Btu/year (which is roughly 30×10^{12} W). A major advantage of the ocean thermal technique is that the ocean serves as an energy storage system, so that power plants should be able to run day and night.

Wind Energy

Up until the middle of the nineteenth century, wind energy played a significant role in the U.S. energy picture; today it contributes only a minute fraction (<1%) of our total energy demands (3). Wind energy conversion systems (WECS) are basically an indirect utilization of solar energy. Roughly 2% of all solar radiation received by the earth is converted into wind energy in the atmosphere. Estimates have shown that the total capacity of the winds surrounding the earth is on the order of 10^{11} GW (gigawatts). The 1972 Solar Energy Panel of the National Science Foundation and NASA estimated that the potential power available from surface winds over the continental United States, from the eastern seaboard to the Aleutian Arc, is equivalent to about 10^5 GW of electricity. This represents roughly 30 times the estimated total U.S. power consumption at present (3).

The conversion of solar energy to wind energy occurs at all levels; however, 30% of the wind energy is produced in the lowest 3280 ft of the atmosphere (51). Hence, only a small portion of the energy flux in the lower levels is available for use.

There are a large number of favorable geographical locations in the United States where windpower conversion is feasible. Areas showing great promise include regions along the coastal margins and throughout the Great Plains from Texas to the Dakotas. In addition, ocean-based windrotor systems have been suggested for harnessing steady offshore winds (3).

The potential performance range for a WECS is relatively large as it is based on an exponential relationship between wind velocity and output. This gives a high incentive to locating systems at sites experiencing steady high winds. For a conventional windmill, the output from its rotor is directly proportional to the square of the blade diameter and the cube of the wind velocity.

Although a WECS's output is proportional to the cube of wind velocity, it is usually not economical to design electrical generating systems to absorb 100% of the rotor power at the maximum possible wind speeds. (Figure 2.7 illustrates a typical WECS for electrical energy generation.)

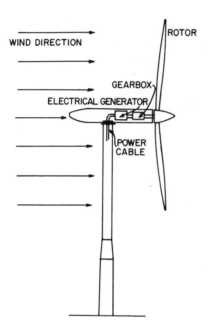

FIG. 2.7 A typical wind energy conversion system (WECS).

As wind speeds are variable, it is generally more cost effective to design smaller generators capable of maintaining a constant output at all speeds in excess of the system-design wind velocity (this is referred to as flat rating).

Wind machines are designed to begin rotating at some minimum wind velocity. Start-up speeds are based on defining a period over which the system will not generate power. A WECS will operate near its design capacity only when the wind is blowing at or above the flat-rate wind speed. During other times, the system is generating power at less than full output. A significant limitation is that a typical WECS will operate at an overall load factor in the range of 15-25%, which is only about one-quarter that of a typical fossil-fueled plant. This means that for a given annual power out-put, a WECS will require roughly four times the installed capacity of a conventional steam plant (3,51).

The theoretical maximum energy that can be harnessed by a WECS is around 60% of the energy in the airstream intercepted by the device's blades. This theoretical recovery is reduced further (to a maximum of about 40%) by blade inefficiencies and mechanical losses. Overall wind efficiency of an individual WECS will probably be no greater than 35%.

A variety of wind conversion schemes have been proposed. For example, the production of hydrogen by electrolysis is being considered where the hydrogen would be employed in an engine or turbine to generate constant-frequency alternating current (AC), or hydrogen could be piped to the user instead of electricity. A number of systems have been proposed for direct current (DC) electricity as well (3).

The future for large-scale windpower usage, if there is one, is almost entirely tied to electrical generation. Central generating systems in the form of large-scale wind farms have been envisioned for supplying the power to existing power networks. Smaller WECS could supply energy to various applications ranging from remote stations to homes.

Because of wind variability, energy storage systems and/or alternate energy sources are essential to WECS to ensure continual supply of energy to the consumer. For the most part, energy storage systems have not been sufficiently developed as yet. Current technology and fuel prices makes storage prohibitively expensive for average homes.

No major adverse environmental impacts have been identified with wind-power usage other than birdkill. No waste heat or air pollution problems are generated. WECS do, however, present some difficulties in airspace, particularly over large wind farms, and interference with radio and televi-sion transmission (3).

The initial capital investment in WECS is expected to be high. Also, since pilot systems have not been demonstrated on a large scale as yet, initial operating costs are expected to be high until such time as the tech-nology is fully accepted.

IMPACT OF ENERGY CONSUMPTION
ON THE BIOSPHERE

In Chap. 1, various chemical cycles were discussed, namely, those of oxygen, carbon, and nitrogen. Also, Fig. 2.2 illustrates the water cycle, or—as it is more commonly called—the hydrologic cycle. Human activities, particularly in the area of energy utilization, have tampered to a certain extent with all of these phenomena. The carbon cycle, more specifically that of CO_2, has been altered significantly since the start of the industrial age.

Since around 1850, human activities have caused marked increase in the amount of CO_2 in the atmosphere. This increase has been from around 290 ppm (parts per million) to roughly 330 ppm, or about a 14% increase in 130 years. Woodwell (53) points out that nearly 25% of this total increase has come within the past decade; at the present rate of CO_2 increase, by the year 2020 concentrations could approach twice the current level.

The CO_2 content in our atmosphere appears to vary seasonally. It reaches a maximum at late winter or early spring (April in the Northern Hemisphere) and a minimum towards late summer or early fall (September or October). Seasonal variations in general appear to be a worldwide trend.

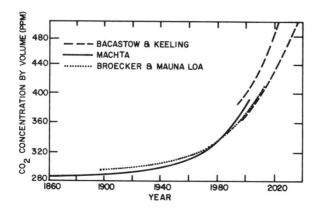

FIG. 2.8 Measured and projected trends of CO_2 concentration in the atmosphere.

As pointed out earlier, the CO_2 level is an important parameter that plays a role in the metabolism of the biota (i.e., the totality of living matter). Seasonal variations in the atmosphere plays a critical role in determining the amount of photosynthesis that takes place during the summer.

Carbon dioxide is a trace gas in the atmosphere, occupying roughly 0.03% by volume; it does, however, play a major role in controlling the earth's climatic conditions, as it absorbs radiant energy at infrared wavelengths. The heat that is absorbed has the potential of altering climatic conditions significantly.

Woodwell (52) believes that seasonal variations in atmospheric CO_2 levels are due to the photosynthetic cycle that occurs during the summer months in the middle latitudes of both hemispheres. He points out that seasonal changes in CO_2 level might be explained by "the pulse of photosynthesis in forests of the middle latitudes" (52). Because of the wide area covered by forests in these latitudes, it seems plausible that photosynthesis is conducted to a greater extent there during the summer than by vegetation in any other region.

The increase of the CO_2 content in the atmosphere has been attributed to the accelerated release of CO_2 through the combustion of fossil fuels. Figure 2.8 illustrates the increase of CO_2 content in the atmosphere since the so-called second industrial revolution of 1860, with projections beyond the year 2000. The current exponential increase in CO_2 concentration stems from our exponential consumption rate of coal, petroleum, and natural gas. More recent findings indicate that CO_2 contributions may also be due to releases from the biota, primarily through the destruction of forests and the oxidation of humus. Traditionally, biota has been viewed of as a sink for atmospheric CO_2; however, a more recent interpretation is that it is a source of released CO_2 (52). The breakdown of carbon in the bio-sphere is as follows (52):

1. The atmosphere contains 700×10^{15} g.
2. The total worldwide biota contains 800×10^{15} g.
3. Organic matter of the soil contains $1000-3000 \times 10^{15}$ g.
4. The oceans contain $40,000 \times 10^{15}$ g.

The largest source of carbon is contained in the oceans and exists mostly in the form of dissolved CO_2. This dissolved gas is part of the carbonate-bicarbonate system discussed in Chap. 1. The second most important source of carbon in the oceans is in the form of dissolved organics (i.e., ocean humus). It is the upper mixed layers of water that effect the most rapid exchange of CO_2 with the atmosphere. At greater depths, the water contains the most carbon. The exchange of CO_2 between the atmosphere and ocean is a relatively slow process. Carbon moves from the atmosphere through the mixed layers into the ocean depths slowly.

The release of CO_2 from biota and the combustion of fossil fuels is believed to have both local and global effects on climate. For the latter, the impact on climatic conditions stems from both thermal and CO_2 pollution.

Over the past million years, the average temperature of the earth has varied naturally by more than 20°C. Roughly 50 million years ago tempera-tures were some 10°C higher than they are today. Glacial periods existed where the average temperature was 10°C cooler and have occurred during at least three periods during prehistoric times (one or more glacial periods occurred around 600 million years ago; the Permocarboniferous glacial age occurred some 300 million years ago; and the most recent in earth's history was about 5 million years ago). The Antarctic ice sheet, containing a mass of ice that is approximately 59 m above sea level, is believed to have attained its present size and volume about 4 million years ago. In the Northern Hemisphere, continental ice sheets first appeared during the Ice Age, roughly 3 million years ago. It is believed that over the last million years, the Arctic Ocean ice cover has never been less than it is at present. In addition, the earth's climate has been characterized by an alteration of glacial and interglacial periods noted in the Northern Hemisphere by varia-tions in the continental ice sheets. The total sea-level change from maxi-mum to zero glaciation occurred at about 210 m. If currently existing glaciers, which occupy only 3% of the earth's surface, were to melt, they would raise the sea level about 70 m.

A rough climatic cycle appears to be around 100,000 years (46). The most recent glacial maximum in the earth's history occurred about 18,000 years ago and ended abruptly about 10,000 years ago. Glaciers retreated at a rate of roughly 1 km/year and reached their current state approximately 7000 years ago. Over the past 100 years, the earth's average temperature increased about 0.5°C until the 1940s when it began decreasing. Since 1940 the average worldwide temperature has decreased about 0.25°C and by more than 1°C above the Arctic Circle (53-58).

Speculation on the cooling trend since 1940 is that this represents the
end of a 10,000-year glacial period and that cooling is likely to continue.
Analysis of cyclic paleoclimatic data tends to suggest that significant cooling
may take place over the next century (46).

Several human activities might account for changes in the CO_2 levels
and consequently the global heat balance:

1. Man has altered the reflectivity and absorptivity of the earth. Cutting
   forests, agricultural practices, building cities, etc., affect the
   albedo, resulting in a net heating of the earth.
2. Irrigation may cause cooling on a local level; however, its ultimate
   effect is a net heating because evaporation of the irrigation water
   absorbs sunlight.
3. Various industrial activities and incineration practices emit particulates into the atmosphere, resulting in a net cooling because of
   reflection and scattering of sunlight back into outer space.
4. The direct release of heat also has an impact. The use of all types
   of stored energy, whether fossil fuels, nuclear, geothermal, or
   whatever, results in a net heating of the earth's surface.
5. Carbon dioxide emissions from the combustion of fossil fuels have
   previously been cited. Atmospheric CO_2 serves as a radiation
   absorber, resulting in a net heating of the earth.

It is not known to what extent global climatic conditions have been altered
since man's energy consumption activities began to accelerate around the
middle of the nineteenth century. Industrial activities may be no more
significant than volcanic activities of the past, which have caused slight
average temperature variations. The total atmospheric content of fine
particles is estimated at present to be around 4×10^7 tons, and of this
roughly 25% can be attributed to human activities.

Figure 2.9 illustrates the climatic system without human influences.
The primary components of this system are the land masses, the oceans
(including sea-ice), and the atmosphere (including wind, clouds, rain, and
snow). As already noted, the main additions due to human activities are
CO_2, dust or aerosols, fluorocarbons, and other pollutants.

Infrared radiation affects the ambient temperature and hence such factors
as the amount of precipitation and evaporation that occur. An excessive
amount of CO_2 in the atmosphere acts like a filter, preventing infrared
radiation from reflecting back into space. The net result is that the earth's
surface becomes warmer and the stratosphere becomes cooler. This can
have a measurable effect on the climate.

The addition of aerosols and particulates to the atmosphere also may
have a detrimental effect on climatic conditions. A large portion of these
pollutants are introduced into the atmosphere directly in the form of soot
or smoke, again from the combustion of fossil fuels and various industrial
processes. In the United States, one of the most common man-made aero-

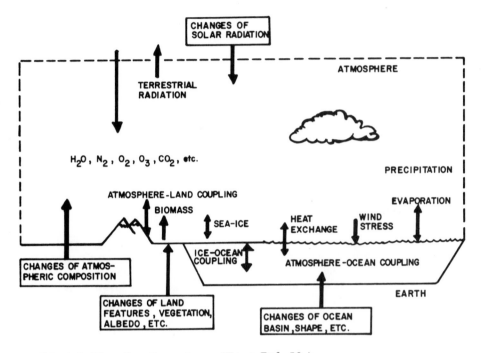

FIG. 2.9 The climatic system. (From Ref. 59.)

sols is sulfate which is formed from sulfur dioxide (SO_2) discharging from the chimneys of coal-burning and fuel-oil-burning power plants. Sulfur dioxide, upon exposure to the air for several hours each day, oxidizes to sulfate. The sulfate readily combines with water, forming droplets of sulfuric acid. These produce the phenomenon called "acid rain."

In general, man-made aerosols are dark and will absorb sunlight. It is therefore believed that when such relatively dark particles exist over land areas they have a heating effect (59). This matter is an area of considerable controversy, as some research results have indicated that aerosols generate a cooling phenomenon and therefore partially explain the cooling trend the earth has experienced since the 1940s (59).

At present, direct power generation is estimated to affect the earth's mean temperature in a minute way (an increase of a hundredth of a degree), although this will certainly change due to the exponential growth in energy consumption shown in Fig. 2.8. The largest contribution to any possible alteration of climatic conditions will come from atmospheric CO_2.

The implications of a warmer worldwide climate are far reaching and will have great impact on future energy planning as well as numerous socio-economic considerations.

Any increase in the release of energy on earth must result in an increase
in ambient temperature. It makes little difference whether the energy source
is fossil fuel or nuclear, or even how efficiently the energy released is
utilized. Essentially, the entire heat produced from any of these forms of
stored energy results ultimately in additional thermal release into the
atmosphere. Geothermal energy may yield slightly less net thermal waste
than fossil or nuclear energy sources; however, only tidal or solar energy
utilization results in no additional release of heat and ensuing temperature
rise. Furthermore, the utilization of solar energy must occur in such a
way that no net additional energy is absorbed by the earth; that is, solar
collectors must not change the albedo of the earth significantly.

SOURCES OF BIOMASS AND BASIC TERMINOLOGY

As already stated, the natural conversion of solar energy into plant materials
via photosynthesis and the further conversion of this stored energy into more
concentrated forms such as petroleum, coal, and natural gas constitutes
the basis of the world's fossil fuel supply. Because these fossil fuels are
being depleted at such a rapid rate, considerable attention is being given
to technologies for providing large additional supplies of high-quality concentrated fuels through the use of forces resembling those which produced
our fossil fuels. The managed production of plant tissue (i.e., from trees,
grasses, water plants, and freshwater or marine algae) with more efficient
use of solar energy and required nutrients carried out on suitable land and
water areas could provide starting organic substances (60).
    Plant matter which constitutes the product of photosynthesis has a com-
paratively low heat content per unit weight. Conversion of plant material
into higher heat content fuels similar in composition and quality to natural
fossil fuel reserves is essential for effective storage and efficient use.
    In natural ecosystems, land plants (e.g., trees or grasses) show a net
productivity in the range of 4-26 tons of dry plant matter per acre per year.
Under conditions of intense cultivation, certain crops can have yields as
high as 40 tons of dry material per acre per year.
    A number of suitable photosynthetic products are generated at varying
yields each year. These major sources of biomass are discussed next.

Forest Residues

    A variety of residues are generated as a result of the growth and harvest-
ing of commercial timber. Included in this class are logging residues,
trees removed in intermediate cuttings, trees and shrubs removed from
the understory in even-aged stands, and the removal of trees killed by
natural causes.
    Commercial logging operations are one of the most important single
sources of forest residues at present. Logging residues are the leftovers

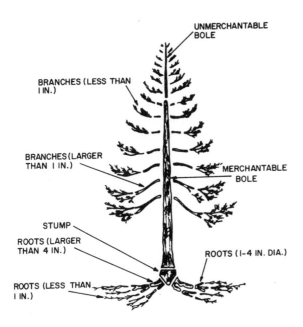

FIG. 2.10 The various components of a softwood tree.

of logging, that is, the material which does not have enough economic value
to justify the expense which must be incurred to remove it.

Most forest products manufacturing operations require only a portion
of the tree, specifically the stem or bole, for use as raw material. Further,
only that portion of the stem which is larger than some minimum diameter,
generally 4 in. or more, is of commercial value (known as the "merchantable bole"). It is the only portion of the tree removed from the forest during
commercial logging operations. The tops, branches, and stump-root systems are not removed during harvesting.

The merchantable stem does represent a large portion of total tree
weight; however, a significant portion of total biomass is found in the other
components. Variability in the proportion of total tree weight represented
by various components is large, and data on the subject are inadequate.
Available data indicate, however, that the merchantable stem generally
represents some 55-70% of total tree biomass, including the stump-root
system, and about 75-90% of total aboveground tree biomass, depending
on tree size, species, and growing conditions (61). Figure 2.10 illustrates
the basic components of a typical softwood tree.

Logging residues, in addition to including tops, branches, foliage,
stumps, and roots of trees of commercial species with usable boles of
merchantable size, include trees of noncommercial species which are

dead and standing or fallen, trees of commercial species which are too
defective to cut for lumber, portions of merchantable boles which are broken
or shattered during tree-felling operations, and finally the understory brush
which grows in many types of forests.

Logging residues exist for only a short period of time during a forest
rotation. This period begins with the timber harvesting and ends when
residues are decayed and incorporated into the soil. The period varies in
length depending on the species involved, and the climate. For example,
in the southeastern United States, logging residues generally persist for
no longer than a year or two, because the wood of many southern species
exhibits only moderate resistance to decay. Pieces left as residues are
relatively small, and the warm, humid climate promotes rapid decay. In
the forests of the Pacific Coast, residues may persist for many years,
particularly when the residues include pieces of very large boles or species
which are highly resistant to decay, such as cedar and redwood.

Logging residues usually must be removed, or at least partially removed,
before a forest stand can be regenerated. This is particularly true following
the logging of large old growth timber on the West Coast, where very large
amounts of logging residue are generated. Residue collection for energy
purposes in such a case would have this additional benefit, the dollar value
of which should enter into any examination of the feasibility of using logging
residues for energy production.

At varying intervals during the life of a commercial forest stand, trees
may be cut in order to increase the amount and value of the timber produced
in the stand. Trees cut for this purpose are referred to as intermediate
cuttings and are usually of poorer quality or smaller size than those remaining (62). Such intermediate cuttings are generally too small to have commercial value.

In all parts of the United States a variety of shade-tolerant shrubs and
trees typically grows beneath the canopies of commercial forests. This
vegetation competes with the commercial species and impedes their growth.
Such competition can be particularly serious in parts of the South.

When understory growth is so vigorous as to present a serious impedi-
ment to tree growth, it is removed—either by burning or by the application
of herbicides. Burning is the traditional method commonly used in southern
pine farms. The southern pines themselves are extremely fire resistant
and can survive a fire which destroys understory brush and trees. Understory brush consists primarily of hardwood species which will usually
sprout again from root stocks after being burned. Hence, understory
vegetation often grows back completely within 1 or 2 years after burning.
Only burning for several successive seasons will destroy root stocks.

As in the case of intermediate cuttings, the removal of understory vegetation should benefit the residual stand of trees as well as provide raw material for energy production. Both benefits should be taken into account in
determining the economic feasibility of such removals.

Large numbers of commercial forest trees die each year from natural
causes, including disease, insect attack, fire, and storms. It has been
suggested that such trees could be salvaged and used for energy production.
The primary problem in this instance is one of collection. Trees killed
by natural agents (e.g., spruce budworm in the Northeast), are generally
widely dispersed. Hence, collection of most of this material would not be
economical, except in areas where commercial logging is in progress. In
this latter instance, mortality usually becomes part of logging residues.

There are cases where mortality is concentrated in certain areas.
Examples of this would be large acreages of timber killed by the tussock
moth in the Pacific Northwest. In these instances attempts are generally
made to salvage the timber for use in primary products before decay becomes serious. Logging residues would also be generated by these operations. It is unlikely that concentrated areas of mortality would otherwise
be a feasible source of material for energy production. Much of the salvage-
able material is suitable for competing uses; also, such concentrations of
mortality are random occurrences and do not take place with enough regular-
ity in any region to provide a continuous supply. Other examples of such
occurrences are pine beetle kills in the South and the Pacific Northwest.

The important characteristics of forest residues with respect to energy
production are heating value, ultimate analysis, moisture content, particle
size, and bulk density. Forest residues differ from mill residues primarily
in particle size and bulk density, and—perhaps most important—in the
admixture of wood, bark, and foliage that is characteristic of forest residues. Wood and bark residues are generated at separate locations in a
paper mill and may be separately collected if so desired. With present
technology it does not appear feasible to debark a large proportion of forest
residues, particularly branches, which are relatively small and irregularly
shaped.

In addition to wood and bark, forest residues may contain some proportion of foliage, depending on which residues are collected, the method of
collection, and the elapsed time between residue generation and collection.
For example, the collection of logging residues following harvesting of an
old growth Douglas fir stand in the Pacific Northwest can be accomplished
with conventional machinery. However, such collection would involve only
relatively large residue material, as it is usually not feasible to remove
branches and small trees by such methods. Thus, any residues collected
would probably contain little or no foliage (62). Figure 2.11 shows one
type of brush-clearing machinery used.

Residues from thinnings would probably contain relatively large amounts
of foliage, since such residues would consist of whole trees (excluding
roots). Data collected by Young (63) indicate that foliage represents about
37% of the aboveground dry weight of 10-ft-tall red spruce trees. This
percentage declines with increasing height, to 12% for 40-ft trees.

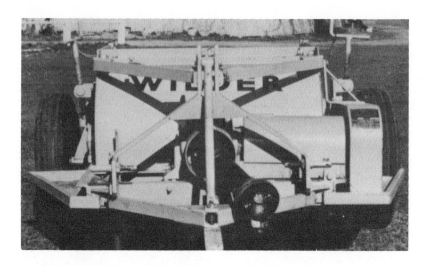

FIG. 2.11 A heavy-duty forestry clearing machine from Britain clears
scrub, brush up to 4 in. in diameter, and tree prunings, and leaves the
ground ready for subsequent planting. Designed to be towed behind and
operated from a standard tractor, the Scrub-Masta can be easily maneuvered
between trees and other obstacles to clear strips up to 40 in. wide. It was
built to meet Britain's Forestry Commission requirements, will tackle
scrub up to 20 in. high, and will cope with branches up to 4 in. diameter.
The vertical cutting and mulching action is performed by 26 flails mounted
in two banks which are guarded by a steel canopy to prevent pieces of broken
or cut material from being thrown about. Four types of flail are available
for quick interchange to suit different terrain and different kinds of scrub:
a "blockbuster" for clearing heavy scrub; a "straight" for pulverizing brush
and tree prunings; a "cranked" for clearing lighter scrub; and a "silage"
for clearing lightweight twiggy growth. The vertical action of the flails
permits use of the machine in areas where tree stumps, stones, and out-
croppings are encountered. (Courtesy of John Wilder Engineering, Ltd.,
England.)

    The proportion of bark in forest residues would depend primarily upon
the size of forest residues collected. As a general rule, the weight of
bark as a percentage of total stem or branch weight varies inversely with
stem or branch diameter (63,64).
    The woods of softwood species exhibit heating values ranging from 8200
to 9700 Btu/lb, with resinous woods exhibiting the higher values. The hardwood species generally exhibit heating values ranging from 7900 to 8800
Btu/lb, with most species near the lower end of that range.

Bark generally exhibits higher heating values than wood, ranging from 8280 to 10,260 Btu/lb for several softwood species and from 7070 to 9490 Btu/lb for hardwood species.

Foliage may be an important component of forest residues also. Little information is available on the heating values of foliage, particularly hardwood foliage. However, this should not pose a major problem in planning for energy utilization of forest residues. The range of heating values for softwood foliage has been reported in the range of 7999 to 9050 Btu/lb, which is not very different from the range of heating values for the woods of softwood species (62).

No data are available on the heating values of hardwood forest tree foliage specifically, but Golley (65) has reported an average value of 7612 Btu/lb for the leaves of 57 plants. This value is slightly lower than the heating value of woods of most hardwood species but considerably higher than the assumed heating value of agricultural crops used by Szego and Kemp (66) in their examination of the energy farm concept. Heating values of forest residues will vary according to the species involved and proportions of wood, bark, and foliage included.

All forest residues contain some moisture. The moisture content will vary greatly depending on the species involved, the length of time residues remain uncollected, and the weather patterns which prevail while residues remain on the forest floor. Fresh residues will exhibit green moisture contents typical of the species involved, whereas residues which are left on the forest floor may dry to some extent, depending on the weather.

The moisture content of fresh residues is in the range of 40-60%, according to Hawley and Smith (62). This study presents green moisture contents for the various parts of several forest tree species. Resch (67) has estimated the average moisture content of forest residues at 30-60%.

Such high moisture content effectively reduces the amount of heat recoverable from residue fuels as heat is required to drive off the moisture as steam and to heat it to flue gas temperature levels. Figure 2.12 illustrates the effects of various moisture contents on the combustion efficiencies of wood and bark, respectively (68).

Another important parameter is residue material size, which before processing varies from small twigs and needles to whole logs several yards in length. All recoverable residues can, however, be processed through a chipper.

Chipping machines currently available to the pulp and paper industry produce chips ranging up to 1 in. in length, 3/4 in. in width, and 1/4 in. in thickness (69, 70). When pulpwood is processed with such a device, the resulting chips generally contain 2-5% of dust-sized particles, by volume, and up to 3% of coarse material, consisting of slivers which may be up to several inches in length (70). Figures 2.13 and 2.14 show two types of chippers.

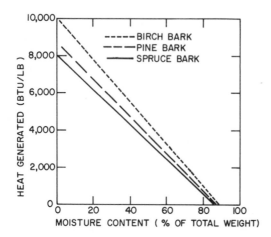

FIG. 2.12 Effects of moisture content on heat generated by bark combustion.
(From Ref. 68.)

Use of standard chippers to process forest residues would likely generate
similar proportions of fine and coarse (>1 in.) materials. However, the
proportion of particles between dust size and full chip size could be significantly greater, depending on the proportions of foliage, small branches,
and twigs included in collected residues.

Detailed information on the quantities of forest residues is limited.
Quantities reported later in this section are estimates based on the best
available data. Forests, like other biological communities, may vary
widely within relatively limited areas.

Logging residues consist of leftover material from growing stock, from
non-growing-stock trees, and from trees of noncommercial species growing
on commercial forest lands. Growing stock, as defined by the U.S. Forest
Service, includes only "live trees of commercial species qualifying as
desirable or acceptable trees." (71). Excluded from this definition, besides
trees of noncommercial species, are trees of commercial species which
are "rough, rotten and dead."

The only consistent and complete estimates of logging residues on a
nationwide basis are made by the U.S. Forest Service. These estimates
include only residues from so-called growing stock volumes. Growing-
stock volume, as defined by the Forest Service, refers to wood in the stems
of growing stock trees which are 5 in. or more in diameter at breast height
(4-1/2 ft above ground level). Volume is calculated from a 12-in. stump
height to a minimum 4-in. top diameter.

Table 2.1 lists total logging residues by region for the year 1970. As
far as sheer volume of logging residue production is concerned, the Pacific

FIG. 2.13 Vertical fed chipper for chipping short waste such as trim blocks, furniture, waste flooring, and waste and short slabs. (Courtesy of Fulghum Industries, Inc.)

FIG. 2.14 Chipper with common base for chipper and electric motor. (Courtesy of Fulghum Industries, Inc.)

TABLE 2.1 Total Logging Residues by Region in 1970

Region[a]	Residues from Growing-Stock Volume[b] (10^3 DTE)	Total Residues[c] (10^3 DTE)[d]
New England	1,408	3,976
Middle Atlantic	2,259	5,331
Lake States	865	3,670
Central States	1,405	4,530
South Atlantic	4,935	11,813
East Gulf	2,074	5,622
Central Gulf	3,465	10,585
West Gulf	3,471	9,694
Pacific Northwest	7,715	18,361
Pacific Southwest	1,970	4,937
Northern Rocky Mountains	1,535	3,600
Southern Rocky Mountains	345	1,079
Total United States	31,447	83,198

[a]Regions defined as follows:
  New England—Maine, New Hampshire, Vermont, Massachusetts, Connecticut, Rhode Island
  Middle Atlantic—Delaware, Maryland, New Jersey, New York, Pennsylvania, West Virginia
  Lake States—Michigan, Minnesota, North Dakota, South Dakota (east), Wisconsin
  Central States—Illinois, Indiana, Iowa, Kansas, Kentucky, Missouri, Nebraska, Ohio
  South Atlantic—North Carolina, South Carolina, Virginia
  East Gulf—Florida, Georgia
  Central Gulf—Alabama, Mississippi, Tennessee
  West Gulf—Arkansas, Louisiana, Oklahoma, Texas
  Pacific Northwest—Alaska (coastal), Oregon, Washington
  Pacific Southwest—California, Hawaii
  Northern Rocky Mountain—Idaho, Montana, South Dakota (west), Wyoming
  Southern Rocky Mountain—Arizona, Colorado, New Mexico, Nevada, Utah
[b]Growing stock includes live trees of commercial species qualifying as desirable or acceptable trees. Growing-stock volume is the net volume of the stems of growing stock trees 5 in. or more in diameter at breast height ($4\frac{1}{2}$ ft above ground level), from a 12-in.-high stump to a minimum 4-in. top diameter.
[c]Total residues include residues from growing-stock volume, residues from non-growing-stock volume, and tops and branches. Not included are trees and shrubs of noncommercial species, regardless of size, trees of commercial species less than 5 in. in diameter at breast height, and stump-root systems.
[d]DTE, dry ton equivalents.
Source: From Ref. 61.

Northwest leads the nation. Total logging residues in the Pacific Northwest amounted to 18.4 million DTE (dry ton equivalents) in 1970, or 22% of the national total of 83.2 million DTE. If residues from growing stock alone are considered, the Pacific Northwest in 1970 produced 7.7 million DTE, or 25% of the national total of 31.4 million DTE.

In the case of the Pacific Northwest, some important information is lost in using the average for the region. It is in the western areas of Oregon and Washington, the portions of those states west of the summit of the Cascade Range, that the old-growth Douglas fir forests are found. The old-growth stands are generally more than 150 years old and, therefore, usually decadent. Hence, they yield enormous quantities of both usable timber and residues when harvested. The normal method of harvesting is by clearcutting.

From an economic standpoint, not all residues left in the forest after logging are available. Estimates of residues from growing-stock volumes and from non-growing-stock volumes are good, conservative approximations of logging residues that, because of their size, may be recovered by using relatively conventional methods. Estimates of total residues are likewise conservative, because of the several sources of residue which are omitted. However, the nature of the materials that make up the difference between residues from growing-stock and non-growing-stock volumes and total residue is such that new equipment and methods may have to be devised before such material becomes economically available as fuel.

It might indeed be economically feasible to recover a portion of the stump-root systems in some parts of the South. However, cost data are scanty and cost estimates for a full-scale operation are based only on conjecture. Probably this is something that will never be feasible in many areas because of problems of terrain, tree size, or the attendant soil disturbance, which many silviculturalists find unacceptable.

At present, intermediate cuttings are not an important source of forest residues. Estimates indicate that timber stand improvement practices, which consist primarily of intermediate cuttings, were carried out on an average of 1,413,000 acres/year in the 1968-1971 period (71). This is about one-third of 1% of the commercial timberland in the United States. It is extremely difficult to obtain estimates on the biomass of residues currently generated by intermediate cuttings, because of the great variety of situations in which intermediate cuttings are used. Most kinds of intermediate cuttings are not carried out regularly, but on a when-needed and where-needed basis in order to correct defects in a growing stand of trees. Residues generated are often in small amounts and widely and randomly scattered.

Understory removals in southern pine farms are currently accomplished by using herbicides and controlled burning. Burning is the cheapest method and probably the most commonly used. Unfortunately, no data are available on the amount of underbrush killed by these methods each year or even on the number of acres treated.

Based on the growth rates of many hardwood species, a reasonable
estimate of understory growth would be something on the order of 0.9
DTE/acre per year. Thus, a 100,000 acre pine farm in an area where
understory growth is a problem might produce around 90,000 DTE/year
in understory growth. There are presently no methods for collecting this
material (61).

Understory growth diverts water and nutrients from crop trees to non-
economic trees and shrubs, thus imposing an economic loss as far as crop
growth is concerned. Mechanical removal does not eliminate this problem
but merely postpones it, since the unproductive vegetation will sprout again
each year from uninjured root systems. Herbicide applications or several
successive years of controlled burns, on the other hand, usually destroy
these root systems, solving the understory growth problem for several
years and making the productivity of the site available for the growth of
the crop trees. This factor should be considered quantitatively in any proposed program of mechanical understory removal as opposed to more
conventional methods.

Estimates show that about 4.5×10^6 ft^3 of growing stock volume in the
United States was killed by natural causes in 1970, which represents about
0.7% of total growing stock volume (61, 71).

Tree mortality in 1970 represented, in terms of biomass including bark
and branches as well as growing-stock volume, about 99 million DTE. As
previously discussed, however, most of this material is widely and randomly
scattered, and probably does not offer much promise of economical collection. About 250 million ft^3 of dead timber was salvaged in 1970. Most of
this was concentrated in specific areas, such as timber downed by hurri-
canes and timber killed by major insect infestations.

Water Plants

Floating water plants, such as water hyacinth, are nuisance plants in
many rivers and lakes in tropical and semitropical regions. Their rate of
growth is rapid, and in nutrient-rich bodies of water net productivity of up
to 85 tons of dry product per acre per year has been reported (72). Methods
presently employed in the harvesting of these plants are not highly sophis-
ticated.

There are a number of technical problems associated with the growth
and harvesting of floating water plants. Some of these include predation
by animals and the elimination of nuisance insects; furthermore, there is
need to develop innovative harvesting and transport procedures in order to
minimize energy expenditure.

Algae

Considerable interest is being generated in the use of blue-green algae
in energy-converting technologies. Since these photosynthetic micro-

organisms are capable of trapping and storing solar energy as cell biomass, they can be anaerobically digested to produce methane gas, or can act as catalysts, assisting in the biophotolytic production of hydrogen from water.

The growth of algae for the purpose of removing nutrients in sewage oxidation ponds is common practice. Recent studies (73, 74) indicate continuing yields of 20-30 tons of dry material per acre per year with peak production at a rate of approximately 50 tons/acre per year.

Organic Wastes

The principal sources of biomass in this category are agricultural plant wastes, animal wastes, and urban solid wastes.

Among agricultural plant wastes are included crop residues left in the field at the time of harvest. The principal problem associated with using most crop residues is that of material collection and transportation. For example, although the estimated annual amount of wasted straw is more than enough to satisfy all of the annual U.S. cellulose demand, it is not used because of economic considerations (60).

Estimates in the early 1970s indicate that there were roughly 118 million head of cattle being raised in the United States (75, 76). Based on these estimates, roughly 13×10^7 tons/year of dry organic wastes is generated. Probably the largest portion of this waste is found on open range and is not recoverable. If only 10-20% of the U.S. cattle population are maintained in restricted area feed lots, the total amount of organic waste produced is in the range of 13×10^6 to 26×10^6 tons/year of dry organic matter (60). Smaller amounts of wastes also are generated from other animals restricted to confined areas. It should be noted that local concentrations of these wastes could be quite large.

The average urban waste generated per person is estimated to be 5 lb/day (77). Of this amount, roughly 50% is dry organic matter consisting of waste paper, kitchen wastes, and garden and/or lawn wastes. This amounts to approximately 10^8 tons/year of dry organic material for the entire U.S. population. Cities in general experience severe solid waste disposal problems, and these figures are in the range of 2×10^7 to 4×10^7 tons/year.

The total amount of wastes available for conversion to fuel or energy directly from animal and urban sources alone might amount to 4×10^7 to 6×10^7 tons/year of dry organic material, assuming prohibitive collection costs do not exist. The average heat content of this material has been estimated to be 16×10^6 Btu/ton, which translates to an annual energy supply of 0.6×10^{15} to 1.0×10^{15} Btu. This is roughly 6% of the current energy demands of electrical generating plants in the United States (60).

The technical problems associated with the use of organic waste material as sources of energy largely stem from the variable composition and tendency towards degradation of wastes. These problems arise during storage and transport of unprocessed wastes.

3

Socioeconomic and Environmental Factors

BIOMASS PRODUCTION SCHEMES: CONCEPTUAL MODELS

Biomass production and conversion systems employ various technologies and processes to transform feedstock into energy-related products. Structured biomass production schemes being considered include:

1. Silvicultural biomass farms
2. Agricultural biomass farms
3. Forestry residues
4. Crop residues

Some of these systems are purely conceptual in design, whereas others are considered accepted technology.

Figure 3.1 illustrates the various segments of biomass production, conversion, and marketability. The U.S. Department of Energy [DOE— incorporating the Energy Research and Development Administration (ERDA)] has indicated the following to be the most viable sources of biomass for fuel conversion:

1. Terrestrial biomass—primarily hardwood trees cultivated on energy farms.
2. Sugarcane—in some regions of the world (for example, Hawaii, Puerto Rico, and some southern states), significant amounts are produced; however, sugarcane represents only about 5% of the potential offered by woody species.
3. Residues from lumbering, field crops, and animals—these represent a small fraction of biomass; however, they will be used in the short-term future because of their availability or until a long-term supply of biomass is assured by establishing energy farms.

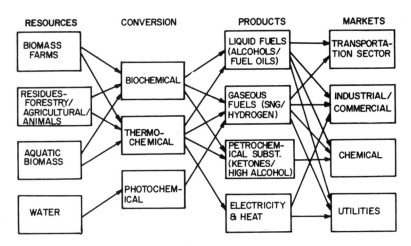

FIG. 3.1 The flow of biomass from raw feedstock to fuel products to various markets. (From Ref. 78.)

4. Accumulated standing forest vegetation not currently harvested.
5. Aquatic biomass—developing technology for harvesting and using aquatic plants for energy is considered to be a long-range research project; the large areas potentially available in the oceans justify development of this technology.

Two attractive features of biomass fuel are its widespread availability and the fact that it is a renewable resource. Expanded use of crop and forestry residues is possible at present. The resource base might be expanded by establishing biomass farms that are designed specifically for the production of fuel fiber. Two limiting factors which might hinder this development are feedstock cost and relatively low energy density (i.e., the ratio of fuel generated to land utilized). Discussions of the various approaches to mass production of biomass feedstock and the issue of land availability are presented in the following sections.

SILVICULTURE BIOMASS FARMS

No silvicultural energy farms are in existence at present. Szego and Kemp (79, 80) discuss the concept in detail, and evaluations are given by Alich and Inman et al. (82).
The major sources of wood biomass which offer current or future potential for energy production include:

1. Mill residues generated at primary wood manufacturing plants
2. Logging residues generated during the harvest of commercial timber

3. Natural stands of surplus or stagnant timber
4. Precommercial and commercial thinnings from commercial timberland
5. Envisioned silvicultural energy farms

Depending on future energy demands and the proximity of users to sources
in various regions of the United States, one or more of these sources can
be expected to play a role in this country's attempt to increase its level of
energy self-sufficiency (82).

Blake and Salo (83) outlined the characteristics of the model biomass
farm: (a) farms should be as large as possible, consisting of up to 50,000
or 60,000 acres of land per farm; (b) production should be aimed at growing
closely planted hardwood trees on a short-term basis, i.e., less than 15
years; (c) trees should be planted in a staggered fashion which corresponds
to the length of time that the trees are allowed to grow; (d) the approach
should consist of utilizing intensive management practices such as irrigation,
fertilization, and pest control to optimize yields; (e) advanced harvesting
systems should be employed to collect biomass in as short a time as possible
during the fall, winter, and early spring; and (f) the farm should rely on
sprouting of cut stumps (regeneration of coppice) for the establishment of
all but the initial seedling or cutting crop. Such farms will probably be
established on sites where the precipitation is at least 25 in./year, the
slope is less than 30% (17°), and the land is currently used as forest or
pasture/range or is used to produce forage crops such as hay as opposed
to field crops such as corn (83).

The sequence which would be followed in producing and delivering a
biomass crop to a conversion facility is illustrated in Fig. 3.2. Site
preparation and establishment would not be anticipated after the initial
harvest because of regeneration by sprouting from cut stumps.

The ultimate design(s) for silvicultural biomass farms will depend on
the specific site requirements and technological developments in biomass
crop management, harvest, and transport.

As noted earlier, such farms might consist of plantings of selected
fast-growing hardwood species at close spacings, with irrigation, fertilization, and pest control as required, harvest of the crop at short rotation
intervals, and delivery of the harvested biomass to a conversion facility
for processing.

The model design for a silviculture biomass farm include a number of
basic features. Each of these is considered in the following paragraphs.

Annual Production. The primary feature of any farm is its level of
production, i.e., the quantity of biomass it is designed to produce each year.

Annual production is a function of acreage planted and the average annual
productivity (DTE* per acre per year) or mean annual increment obtained

---

*DTE, dry ton equivalents.

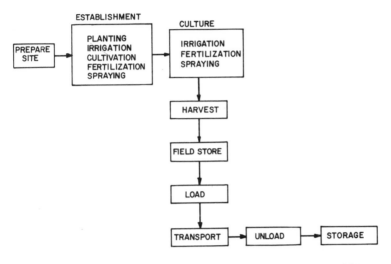

FIG. 3.2 Scheme for the production of fuel on silvicultural biomass farms. (From Ref. 83.)

over the rotation period. The mean annual increment is a site-dependent variable controlled by both the species that can be grown at a given site and the climatic conditions; the size of a farm necessary to provide a chosen level of production is also site dependent.

Biomass Species. The biomass species should consist of trees known for rapid early growth and strong coppicing capabilities. Various species of hardwood trees fit this description.

Rotation Period. A harvest rotation period of 6 years was selected as representative of possible rotation schedules that may be used in farm production in a study conducted by MITRE Corp. (83). A farm production lifetime of 30 years is realistic largely in that it conforms with the expected lifetime of a conversion facility. A 30-year production life and a 6-year rotation schedule accommodates the harvest of five rotations—the first rotation consisting of the seedling or cutting crop, and the subsequent four rotations of coppice crops.

Land Acquisition. Land may be acquired by either direct purchase or long-term lease.

Land Clearing and Preparation. Land acquired for production will likely consist of a mixture of wooded and open land. The existing vegetation on wooded land must be cleared away prior to planting the first rotation. Merchantable timber can be sold to recover the costs of land preparation. The

clearing and preparation of wooded land is more expensive than that for
open land such as pasture and previously cultivated fields. Land clearing
and preparation can be considered as a front-end operation, since it is not
required between harvests if coppicing species are used and so need be
performed but once during the lifetime of the plantation (82,83).

Planting. The biomass species will be planted as seedlings or cuttings
in rows at a specified spacing and planting density. It may eventually be
possible to use seeds for some species rather than seedlings or cuttings.
Planting costs include both the costs of the planting stock and the costs of
planting. Planting is also considered to be a front-end operation, being
performed only once during the plantation lifetime if coppicing species are
employed.

Crop Management. Effective crop management is recognized as playing
the dominant role in increasing mean annual growth increments to economically feasible levels.

Harvesting. The harvesting operation can be performed by some type
of self-propelled biomass harvester, i.e., one which cuts and chips the
biomass and delivers it to a self-dump forage wagon pulled behind the
harvester (Fig. 3.3). The chipped biomass can be stored in the field at
designated storage areas until transported to the conversion facility.
Transport from the harvester to the storage areas can be accomplished
by tractor/forage wagon teams (82).

Roads. Roads are necessary both to allow access to the farm to accommodate crop management and harvesting operations and to supplement existing public road networks for transport of the biomass to the conversion
facility.

Transportation. Transport of the harvested biomass to the conversion
plant might be accomplished by semitrailer trucks with hydraulic dump
capability. The trucks can be loaded at the field storage areas by front-end
loaders.

Field Support. The field support operation serves to transport manpower
and materials within the plantation area, to supply fuel for field equipment,
and to move harvesters and other field equipment from place to place within
the farm as required.

Planning and Supervision. This is a front-end operation, the costs of
which are assigned to the first year of farm life. The supervision component
is an annual cost item, consisting of salaries of office personnel and field
foremen, overhead, and vehicle capital and operating costs.

FIG. 3.3 Forest clearing machine that picks up the entire tree and reduces it to wood chips in less than a minute. (Courtesy of Morbark Industries, Inc.)

Biomass Production Costs. The costs of equipment, materials, labor, and fuel are primary determinants in developing a productive system. Costs of land, fertilizers, diesel fuel, labor, custom operator fees, etc., can be expected to vary from site to site. For each of the component operations such as irrigation, harvesting, and transportation, a number of options are available for consideration. In certain operations such as irrigation, the method of choice may vary depending on differences in physical characteristics between particular sites.

Model Design. Inman et al. (82) selected 10 sites having varying climatic, topographic, and land-use characteristics. A cost analysis was performed for producing silvicultural biomass on intensively managed short-rotation farms at each of the sites. An energy budget for biomass production was also examined. As an example, Fig. 3.4 illustrates the field operations schedule for a model system proposed in this study. In this conceptual scheme, land is assumed to be acquired either by lease or purchase as needed for planting. It is assumed that under a 6-year rotation schedule, one-sixth of the total acreage to be planted is acquired during each of the first 6 years of a 30-year period. Thus, land acquisition is spread over a 6-year period, which corresponds to the first rotation of the first planting. Land clearing and preparation operations are spread over the first rotation period, with land being prepared for planting as it is acquired, as

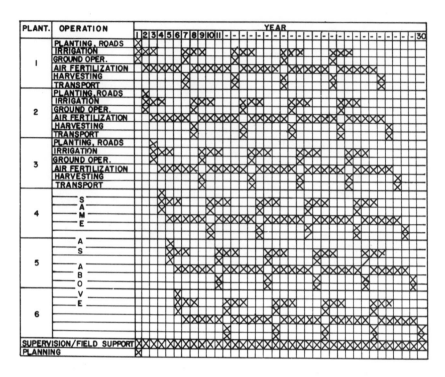

FIG. 3.4 Operations schedule for a conceptual plantation design proposed
by Inman et al. (82) over a 30-year planting life and 6-year rotation period.

proposed by Inman et al. (82). These operations are performed under contract by custom operators. No operators or equipment are specifically
employed or purchased internally by the farm's management for the proposed
operation. Planting occurs in conjunction with land acquisition and clearing
operations, and is also performed by custom operators. Machine planting
techniques for both cuttings and seedlings are well developed. The stagger-
ing of the planting operation over the first rotation period affords the harvest
of the biomass crop each year, starting at the conclusion of the first rotation
period. Under a 6-year rotation schedule, the first 6 years are devoted to
staggered crop planting. The seventh through the thirty-sixth year involves
harvesting of the staggered plantings. Although the life of a given planting
is set at 30 years, the staggered nature of the plantings extends the total
operation to 36 years (82).

Irrigation is accomplished in the first 3 years of each rotation. Irrigation is carried out over a 4-month period during the growing season in this
model. Tractors are required to move the traveler unit from one set to
another. Since the irrigation and harvesting seasons do not overlap, the

same tractors used for the irrigation operation are also available for harvesting. Irrigation field work is conducted by custom operators. Since irrigation is conducted only during the first 3 years of each 6-year rotation period, only half of the total planted acreage is irrigated during any given year between the third and thirtieth years.

Nitrogen, phosphorus, and potassium fertilizers are applied by ground equipment only during the first year of each rotation. Applications during the second through the sixth year of each rotation are made by fixed-wing aircraft. From the sixth year through the thirtieth year, one-sixth of the total planted acreage is fertilized annually by ground equipment, and five-sixths by aircraft. Liming and weed control are also practiced during the first year of each rotation. Mechanical weed control is conducted twice during each treatment year.

Each staggered planting is harvested following the sixth growing season of its rotation period. The harvest season is restricted to the winter months and varies in length from 4 to 5 months, depending on the length of the growing season, during which the entire harvesting operation for the year is completed. Tractor-wagon teams work concurrently with the self-propelled harvesters, exchanging empty wagons with the wagons filled by the harvesters, transporting the loaded wagons for dumping at designated field storage areas strategically located over the farmed area, and returning the emptied wagons to the harvesters.

In the scenario presented by Inman et al. (82) the field-stored biomass is transported from the storage areas to the conversion facility by means of semitrailer trucks. The trucks are loaded at the storage areas by front-end loaders. The biomass harvested and stored during the 4- to 5-month harvesting season is transported over a 12-month period. The harvesting season runs concurrently with the first 4 or 5 months of the trucking season. The length of time that the biomass remains in storage piles can vary from a minimum of a few days to a maximum of some 7 or 8 months.

Field support is a year-round operation. Loaders working the storage piles require servicing throughout the year. During the growing season, most of the support activities are directed toward servicing crop management activities (irrigation, fertilization, liming, and weed control). Harvesters and tractor-wagon teams require servicing only during the winter season. Supervision is also a year-round operation, although the greatest work load may be expected to occur during the spring and summer months in conjunction with crop management.

Silvicultural energy farms will probably be sized to produce a specified amount of biomass. It should be noted that a large amount of land is required to produce a relatively small amount of feedstock, in terms of current conversion plant capacities. A given farm would probably operate under a contract to supply a single feedstock customer, who in turn requires a specified and continuous amount of feedstock to maintain his desired level of output. The biomass supplier would be required to size his operation

FIG. 3.5 Potential biomass yields for silvicultural farms assuming that 10% of forest, pasture/range, and forage cropland are utilized. Values given are in 10^6 dry tons. (From Ref. 83.)

accordingly, within the limits imposed by his ability to acquire land. Depending upon the size of the conversion facility, more than one farm might be needed to supply adequate feedstock to meet the demands of given customers. On this basis, a conceptual farm design requires that the annual level of production be specified.

As an example, for sugarcane biomass, roughly 40,000 to 50,000 acres would be required to approach the upper limit for effective plantation management (84). Difficulty may be anticipated in acquiring much more than 50,000 acres for biomass farm development under a single management. At current expected productivity levels for wood biomass (5-12 DTE/acre per year), a 250,000 DTE annual production level would require 20,000 to 50,000 acres, which is within the size range afforded by present acquisition considerations. If this quantity of biomass were to be burned in a wood-fired electric power plant, it would generate 45 to 50 MW, which is a moderate amount of electric power.

Blake and Salo (83) have estimated current and future yields of potential biomass crops for energy production. These estimates are illustrated in Fig. 3.5.

FARMS FOR AGRICULTURAL BIOMASS PRODUCTION

At present, agricultural sites are not used to grow fuel fiber. Moreover, agricultural systems have not yet been developed to optimize the production of biomass. There are some high-yielding feed crops which provide an

indication of current production costs utilizing conventional farm management practices. One example is silage corn. Blake et al. used a silage
corn farm to illustrate the production characteristics of an energy farm.
Other plants also show promise as potential biomass candidates. Midwest
Research Institute (MRI) conducted a systems analysis to evaluate large
numbers of grains and grasses. MRI (78) has listed the following possible
new crops:

1. Kenaf (Hibiscus cannabinus)
2. Roselle (Hibiscus sabdariffa)
3. Giant reed (Arundo donax)
4. Cattail (Typha latifolia)
5. Guayule (Parthenium argentatum)

Wheat, sorghum, oats, alfalfa, and other traditional crops were also
evaluated. Lipinsky et al. at Battelle-Columbus Laboratories (85) conducted
systems analyses for use of sugar cane and corn as biomass crops.

Agricultural biomass farms may have some characteristics in common
with silvicultural biomass farms. The model, or ideal system, might be
based on the following criteria:

1. The farm should be large in land area.
2. Perennial crops should be grown whenever possible.
3. The system should employ intensive management practices.
4. Advanced harvesting techniques should be used to collect biomass.
5. Sprouting should be relied on for reestablishment.

As noted by Blake and Salo (83), agricultural biomass farms should be
located at sites having characteristics similar to those employed for silvicultural farms: (a) minimum precipitation of 25 in./year; (b) slope less
than 17° (30%); and (c) land that is currently used as forest or pasture/range
and can be used to produce forage crops such as hay rather than field crops
(e.g., wheat). Figure 3.6 illustrates various production and transportation
segments of biomass crop utilization.

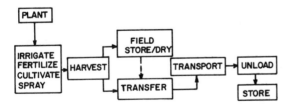

FIG. 3.6 Scheme illustrating one approach to the production of fuel from
an agricultural biomass farm. (From Ref. 83.)

FIG. 3.7 Noncropland regions with high to medium potential for conversion to cropland. Values given as 10^5 acres and percentages. (From Ref. 86.)

Dideriksen et al. (86) indicates that roughly 110 million acres of U.S. noncropland could be converted to crop production. Figure 3.7 illustrates noncropland regions in the United States showing medium to high potential for agricultural biomass production. Almost 40% of this land is in the mountain states and northern and southern plains, where low precipitation is a problem. Erosion is also a limiting factor in these areas of the country; however, a portion of this land could be converted to agricultural biomass production.

AGRICULTURAL RESIDUES

There are approximately 400 million acres of cropland in the United States. Of this amount, roughly 4% are designated cropland pastures. The types of crops produced on this land varies greatly. Under current prices and demands, crops are more valuable as foodstuff or feed than they are as fuel; however they are quickly approaching parity. Harvest residues are even less valuable and many are simply unused. Table 3.1 illustrates the costs of using various agricultural products as fuel sources.

Figure 3.8 illustrates a basic scheme for the collection and transport of agricultural residues to a fuel processing facility. A key step in the

TABLE 3.1 Cost of Using Selected Agricultural Products
as Fuel Sources

Product	1975 Average Price	Estimated Price Per DTE	Estimated Price Per 16 MBtu/DTE
Strawberries	$30.50/cwt[a]	$6100	$380
Wheat	3.52/bu[b]	156	10
Oats	1.43/bu[c]	119	7
Corn for grain	2.46/bu[d]	117	7
Hay, all	49.24/ton[e]	62	4

[a]Hundred weight (pounds), 90% moisture.
[b]60-lb bushel, 25% moisture.
[c]32-lb bushel, 25% moisture.
[d]56-lb bushel, 25% moisture.
[e]Ton, 20% moisture.
Source: Data from Refs. 83 and 87.

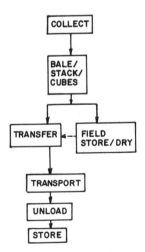

FIG. 3.8 Schematic of collection and transportation activities of agricultural wastes and residues to a conversion facility. (From Ref. 82.)

TABLE 3.2 Average Densities of Residues
from Different Collection Methods

Compaction Form	Weight (lb)	Density (lb/ft^3)
Standard bale	110–150	10–14
Large round bale	1,000–2,500	10–13
Air-packaged rectangular stack	6,000–20,000	4–6
Field cubing	—	16–22

Source: From Ref. 88.

collection process is the compaction of biomass. Residues can be compacted in a variety of ways; the most common are these:

1. Standard bales
2. Large round bales
3. Air-packed rectangular stacks
4. Field cubes

The stack method is generally the least expensive approach; however, it produces the bulkiest package. Table 3.2 gives typical densities for residues packaged according to the various collection alternatives.

Battelle Labs. (89) have estimated transportation costs for crop residues based on 24-ton loads. Some of the estimates from this study are reported in Table 3.3. It should be noted that transportation costs generally increase when loads are smaller, primarily because of the density differences between stacks and bales (90).

In the United States, roughly 322 million DTE of crop residues are generated each year, which on a nationwide basis is about 1 DTE/acre per year (91). Of this amount nearly 86% of the biomass is potentially available, of which 48% consists of small grains and grasses, 35% of grain corn and sorghum (82). Most of this biomass currently has market value as soil conditioners, for erosion control, animal feed, and/or animal bedding and is therefore not considered as a source of fuel. It may, however, be possible to collect 10–15% of the residues produced in various regions without impacting greatly on the aforementioned markets.

Wheat and corn crops represent the most promising source of agricultural residues. Roughly 40% of U.S. cropland is equally divided between these two crops.

Approximately 60% of the currently available residues are produced in the East North Central and the West North Central regions. These represent parts of the country where biomass resources are most likely to be

TABLE 3.3 Estimated Transportation Costs
for Agricultural Residue

Hauling Radius (miles)	Estimated Cost $/Ton (field wt.)	$/Avg. (DTE-mile)	Criteria
0-19.9	1.21	0.28	Transportation costs to supply 1.29×10^6 tons (field wt. with 42.8% dry matter) per year
20-49.9	2.34	0.16	
50-80	3.39	0.12	
0-19.9	1.21	0.15	Transportation costs to supply 0.69×10^6 tons (field wt. with 80% sun-dried matter) per year
20-49.9	2.34	0.08	
50-80	3.39	0.07	

Source: From Ref. 89.

developed. Local factors such as residue density and residue marketability
for other purposes as well as topographic and climatic characteristics will
determine the extent to which this occurs.

Since 1950 the amount of cropland utilized strictly for crops is on the
order of 385 million acres (92). Crop production per acre has increased
by more than 50% over that period. Should the total cropland remain the
same but the increase in productivity be cut in half, residue availability
would be roughly 1.2 times current levels by the year 2020, assuming that
residue availability parallels crop production increases (82).

Residue Market Value

Cost of residue as a fuel feedstock can be highly variable and will depend
on the supply, as well as alternative uses and markets. Blake and Salo
(83) indicate that some crop residues can be purchased for less than $1.00
per million Btu, although most residues will be at least that delivered price
(see Table 3.4).

A substantial data and information bank exists which enables us to perform analyses of biomass farm systems; this is primarily because farming
costs for existing crops are so well documented due to their large-scale
economic importance.

Production costs include labor, materials, and equipment necessary to
prepare the fields and to plant and cultivate the crop. A major portion of
the capital cost may be in irrigation. The cost of installing and operating
irrigation systems of the self-propelled tower and rainbird type averaged

TABLE 3.4 Estimated Costs of Delivered Crop Residues
to Conversion Facility

Operations Phase of Component	Low		Medium		High	
	$/DTE	$/MBtu	$/DTE	$/MBtu	$/DTE	$/MBtu
Residue cost	3.10	0.19	12.50	0.78	30.00	1.88
Collection (large round bales)	7.90	0.49	15.10	0.94	22.30	1.39
Transportation	3.00	0.19	5.00	0.31	8.00	0.50
Total costs	14.00	0.87	32.60	2.03	60.30	3.77

Source: From Ref. 83.

about $30 per acre per year in 1973 (82). This cost includes purchase, installation, operation and depreciation of the entire system.

The purchase price of water is highly variable with different regions of the country. It may range from zero in some areas to more than $20 per acre-foot in others.

Harvest costs include labor, the operation and depreciation of equipment, and short-range hauling of biomass and equipment.

Based on 1973 estimates, at a yield of 15 tons/acre, total biomass production costs are on the order of $12 per ton; at a yield level of 30 tons/acre, such costs are around $18 per ton (51). With these estimates a 20% profit can be realized. In terms of energy production, at 7500 Btu/lb of dry plant tissue, costs would vary from $0.80/MBtu to $1.20/MBtu. For comparison, in 1973 coal sold at around $0.79/MBtu.

Energy plantations probably will involve a low-density operation, characterized by large-scale replication of unit equipment and functions. Economic constraints are likely to be established by total nationwide capacity. This, in turn, should broaden the market and production base for such equipment and so lower capitalization requirements for each plantation.

AQUATIC BIOMASS PRODUCTION

The development of algae as an energy biomass has received attention only recently. Estimates of resource availability are only now being made; however, it is known that the potential supplies of biomass from aquaculture are very large and could substantially contribute to worldwide energy supplies.

Data on projected land-based algae farms indicate that less than 1 Q (quad) of methane could be generated per year at a cost of $2.00/MBtu.

Blake and Salo report (83) that no more than 2 Q could be produced for $3.50/MBtu, and additional quantities would cost up to $6.00/MBtu. Hydrogen generated by means of biophotolysis might cost around $4.00/MBtu if production efficiency were increased by a factor of 10 (93).

Cost estimates on high-Btu gas generated by kelp grown on ocean farms have been reported in the range of $2.30/MBtu to $6.90/MBtu. Budhraja et al. (94) computed a base-case value of $5.00/MBtu assuming an annual yield of 42.5 DTE. (Note that roughly 8.8 MBtu/DTE was assumed for the analysis.)

Aquacultural Farms: Design Concepts

Figure 3.9 is a conceptual design of a plant for the biosolar production of fuels from algae. Some cyanophytes, i.e., filamentous blue-green algae with heterocysts, can synthesize methane and ammonia; such cells can also generate hydrogen gas from water by the process of biophotolysis.

These blue-green algae may be grown in open ponds of water in the presence of waste carbon dioxide and recycled inorganic nutrients. Gas

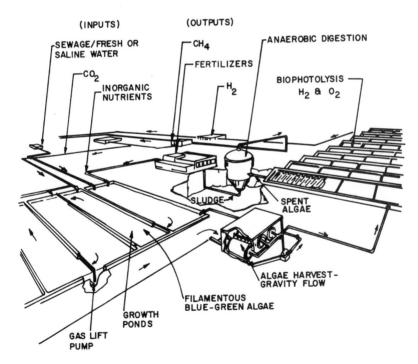

FIG. 3.9 Conceptual design of an aquacultural energy farm.

lift pumps introduce CO_2 to the system and also serve to circulate the growing algae. After an incubation period, algae are collected in a trough by clumping, sedimentation, and/or flotation techniques and then dewatered. The harvested biomass is then deposited in a ciophotolytic reactor where, under a carefully controlled environment, the algae cells use sunlight to split water molecules, forming hydrogen and oxygen. After a specified time period, exhausted algae are either returned to the reservoir for rejuvenation or deposited in an anaerobic digester. In the latter, fermentation takes place and methane and CO_2 gases are formed. A residual sludge consisting of fixed nitrogen remains and is suitable as a fertilizer. Nitrogen can be recovered from the dewatered biomass by pyrolysis, and the inorganic nutrients can be returned to the pond to assist in further biomass growth.

The biophotolysis reactor produces a mixture of gases which must be separated. Oxygen and hydrogen can be separated by diffusion, compression, centrifugation, and/or adsorption. Methane and CO_2 can be separated in the anaerobic digester by an alkali process.

The ultimate design and standard operational procedures for such a plant are far from being finalized. A number of technical problems must first be addressed. These include: (a) locating adequate supplies of CO_2 and water; (b) perfecting and attaining maximal algal growth and harvesting techniques; (c) developing more efficient biophotolysis; (d) establishing optimum conditions for anaerobic digestion of algal growth; and (e) perfecting a process for separating ammonia and various product gases.

The basic concept makes use of solar energy to promote algal growth in outdoor systems. This means that productivity will be a function of regional sunlight and temperature and will vary seasonally. During the summer months, when temperature and light intensity reach their seasonal peak, maximum biomass yield is to be expected. It will be necessary to regulate pond depth and the harvesting schedule in order to optimize biomass yield over longer seasonal periods. Such factors will assist in controlling pond temperatures, thus assuring optimal use of available sunlight (95).

The success of such an approach will depend largely on the availability of CO_2, water, sunlight, and land. These resources must be available on a scale at a cost that is compatible with the market. Both the water and CO_2 supply must first be tested to ascertain whether they are acceptable sources for algal growth. For example, industrial waters and sources of CO_2 often contain significant amounts of toxic materials.

Potential sources that will support algal growth include: (a) water from sewage; (b) some industrial wastewaters; (c) saline groundwater supplies; (d) marine waters; (e) water supplies from geothermal and geopressure wells; and (f) water from agricultural waste runoff.

There are several species of indigenous marine blue-green algae; some of these are <u>Oscillatoria</u>, <u>Trichodesmium</u>, and <u>Nostoc</u> (96). Common freshwater genera include <u>Anabaena</u> and <u>Aphanizomenon</u>. Many freshwater genera are also capable of growing in saline environments. For example, Anabaena and Aphanizomenon have been found in the Baltic Sea. Some algal growth have been identified in hot springs and saline lakes which are enriched with arsenic, copper, lead, nickel, boron, and lithium at seemingly toxic levels (97). Algae obtained from such sources may show good resistance to high concentrations of other toxic compounds.

Inexpensive CO_2 sources might be obtained from sewage and feedlot waste and coal gasification and liquefaction plants. Shale oil recovery generates CO_2 from the partial combustion of oil and the decomposition of carbonaceous rock. Drilled wells and mineral springs are also sources of CO_2, as are fossil-fueled power and industrial plants. The latter also produce daily quantities of sulfur oxides and nitrogen oxides; however, a balanced microbial population would probably be resistant to such oxides. Algal biophotolysis might, for example, provide a cost-effective method of handling pollutants from stack gases (93).

The magnitude of methane generated will depend largely on the source of carbon. Algae cultured on municipal sewage yield under 1×10^{15} Btu/year. Algae grown on animal wastes produce up to 2×10^{15} Btu/year. When algae are grown on waste CO_2, yields in the range of 10×10^{15} to 20×10^{15} Btu/year have been projected (93,98).

Economic Considerations

Jeffries et al. (93) have performed preliminary cost estimates on methane and hydrogen production from algae. The study indicated that algae grown in sewage oxidation ponds would yield relatively low-cost methane. When grown on waste CO_2, methane yield is estimated at a cost slightly higher than projected costs for methane from coal. Methane production from waste CO_2 is estimated to be greater than that from sewage-based sources. Jeffries and associates point out that if coal gasification plants were the primary source of waste CO_2, methane from algae could significantly diversify the base of methane synthesis and reduce the environmental impact of coal use.

The assumptions Jeffries and his coworkers used in their economic analysis were as follows: (a) 6% photosynthetic efficiency for biomass production; (b) inexpensive sources of CO_2 and water; (c) incident solar radiation of 4500 kcal/m^2 per day (or 6.5 MBtu/m^2 per year); and (d) a biomass-to-methane conversion efficiency of 60%. Construction costs in the analysis included those of sewage oxidation ponds built in the central valleys in California at $2.50/$m^2$ and the LLL-Sohio solar thermal pond in New Mexico at $60/$m^2$. Moreover, a systems analysis was carried out of gas-processing facilities for proposed marine energy farms (93,94).

Land costs are a major segment of the total capital investment of sewage oxidation ponds. In 1973, this method was considered to be competitive in comparison to alternative treatment methods, provided land costs were less than $2.50/$m^2$, i.e., $10,000/acre (93).

TABLE 3.5 Estimated Costs of Projected Algal Growth Pond[a]

Item	Animal-Waste-Grown Algae	CO_2-Grown Algae
Construction	3.04	10.13
Accessory structures	4.05	5.27
Sealant	—	0.81
Pipe and gaslift pumps	—	3.04
Subtotal	7.09	19.25
Land	250.00	250.00
Total	257.09	269.25

[a]Costs in dollars per acre.
Source: From Ref. 93.

The cost of algal growth ponds will vary with the nature of the carbon source. Table 3.5 gives the estimated costs generated from the study of construction of ponds used in disposing of animal manure and handling waste industrial CO_2. The estimates are based on 1975 figures, assuming $2.00/m^2$ for handling animal manure and $5.00/m^2$ for CO_2. Note that a substantial difference exists between the two subtotals in Table 3.5. Significantly higher costs are associated with site preparation, sealant, and gas lift pumps when CO_2 is employed.

For methane production, total investment costs include those associated with harvesting and processing as well as pond construction. Total capital investment for a production capacity of 1 MBtu/year for each of the three systems analyzed is as follows (93): (a) for a sewage-based system, $8.60;

TABLE 3.6 Operating Costs and Total Investment
for Methane-Producing Systems[a]

	Sewage-Grown Algae	Animal-Waste-Grown Algae	CO_2-Grown Algae
Operating costs	0.65	1.00	1.60
Total investment	8.60	17.30	30.90

[a]Values given in dollars per MBtu per year.
Source: From Ref. 93.

TABLE 3.7   Estimated Operating Costs for Hydrogen Generation
from Algae[a]

Item	Sewage-Grown Algae	Animal-Waste-Grown Algae	CO_2-Grown Algae
Labor	—	0.08	0.15
Electricity	—	0.02	0.10
Maintenance	—	0.05	0.25
Water	—	0.02	0.40
CO_2	—	—	0.05
Harvest	0.10	0.10	0.10
Processing	0.55	0.55	0.55
Total	0.65	0.82	1.60

[a]Values given in dollars per MBtu.
Source: From Ref. 93.

(b) for a CO_2-based system, $30.90; and (c) for an animal-waste system,
$17.30. Operating costs and total investment are summarized in Table
3.6. These calculations assume the use of solar heating and wind mixing
of the anaerobic digestors. Included are charges from gas compression,
scrubbing operations, storage, and sludge disposal. The analysis showed
that sewage-grown algae would yield gas at roughly $2.00/MBtu; animal-
waste-grown algae, $3.50/MBtu; and CO_2-based systems, $6.00/MBtu.
    Estimated operating expenses for an enclosed hydrogen generation sys-
tem are given in Table 3.7. Jeffries et al. (93) estimated the cost of hydrogen generated by biophotolysis to be on the order of $4/MBtu, assuming
utility financing.

ENVIRONMENTAL CONSIDERATIONS

The environmental impacts to be identified and discussed in this section
are related to activities associated with the production of biomass on silvicultural and agricultural biomass farms and conversion of the biomass to
useful energy forms. Man's agricultural and energy acquisition activities
produce effects beyond the primary goals of food and electricity. Agricultural techniques will be used on silvicultural biomass farms to generate
sources of feedstock that can be transformed into various energy products

TABLE 3.8 Potential Effects of Silvicultural Biomass Farms

Ecological Impact Category Potential Pollutants	Intensity of Impact[a]		
	Site Preparation	Production, Harvesting, and Conversion	Decommissioning
Air quality			
Diesel and gasoline engine emissions	L	M	L
Particulate production	M	L	L
Gaseous emissions from conversion facility	—	M	—
Surface water			
Soil erosion resulting in siltation	H	M	L
Runoff volume increase	M	L	L
Runoff temperature increase	L	—	—
Fertilizer and soil nutrient enrichment of runoff	L	M	—
Herbicide contamination of runoff	—	—	L
Pesticide contamination of runoff	—	L	—
Effluent from conversion and sanitary treatment system	—	M	—
Ground water			
Change in water table depth	L	—	L
Fertilizer accumulation	—	L	—
Pesticide contamination	—	L	—
Ecological impact			
Soil biota change	L	L	L
Aquatic biota change	L	M	—
Changes in wetland areas	L	L	L
Wildlife disruption due to noise, traffic, and gaseous emissions	M	M	L
Wildlife habitat disruption	M	L	—
Vegetational community disruption	M	—	—

[a]Key: H, high; L, low; M, medium.
Source: From Ref. 82.

or fuels. The most significant environmental impacts are likely to stem
from the production and conversion phases. Table 3.8 lists potential
environmental effects that might occur at silvicultural biomass farm sites
along with the intensity of impact as estimated by Inman et al. (82). The
intensity of individual effects will depend on site factors such as soil, climate, slope, the implementation of environmental impact control measures,
as well as the nature of the operational phase. Each of the general topics
in Table 3.8 are briefly discussed in the following subsections.

Air Quality

Air quality can be expected to be affected by vehicle and equipment
engines, as well as particulates (dust) discharged and gaseous emissions
from conversion facilities. Vehicle and utility engine emissions are not,
however, expected to pose serious problems. It can be assumed that all
vehicles used will be equipped with pollution control devices and will be
properly maintained. The vehicles will be operating in rural areas where
background air quality should be reasonably good. During the winter
harvest/trucking season, fewer vehicles should be in operation and therefore no significant air pollution problems should result.

Particulate releases may be a highly objectionable characteristic associated with the transportation (trucks) of biomass from the central collection
areas to the conversion facility. Fugitive dust is generated by pulverization
and abrasion of soil materials by implements (blades, disks, etc.) or by
entrainment caused by air currents produced by vehicles or the wind.

The effect of particulates on human health is usually small; however,
inhalation may aggravate such conditions as asthma or emphysema. Particulates also may be objectionable to residents near unpaved roads experiencing heavy truck traffic. Daily or more frequent water sprinkling of such
roads is one effective dust-control measure. Northern regions of the United
States are likely to require extensive dust-control measures, whereas the
southern sites, which receive more rain, will require less.

Particulates may be more of a problem during site preparation than
during the production phases due to the difficulty of applying control measures. Because of the generally remote locations of such sites, few people
should be impacted. Also, site preparation will only take place during the
first few years of a plantation's lifetime.

Gaseous emissions will be released during conversion facility operations.
While conversion cost estimates include control devices such as bag filters,
precipitators, and scrubbers, low-level releases are likely to occur. Specific emissions will vary and depend on the conversion process that is being
employed. Combustion of the biomass will produce particulates, carbon
monoxide, and minor amounts of sulfur oxides, hydrocarbons, and nitrogen
oxides. Control systems are adequate to mitigate the particulates, and
proper boiler design and operation can reduce carbon monoxide, nitrogen

oxides, and hydrocarbon levels greatly. Sulfur oxide levels should not be significant, as only minor amounts of sulfur are found in wood.

Considerable emission data exist on wood and bark combustion, as this is widely practiced in the paper industry. Currently operated control systems are capable of reducing emissions to within federal primary air quality standards.

On the other hand, pyrolysis and hydrolysis have different and less controllable gaseous emissions. Amines, aromatic hydrocarbons, carbonyl compounds (aldehydes), and phenols may be emitted from these processes. Research is underway to characterize these processes and control such releases. Results should provide better estimates of control methods, costs, and residual releases.

Surface Waters

Surface water impacts may arise from soil erosion, runoff, and increases in water temperature and volume. Runoff may also contain quantities of fertilizer, soil nutrients, and herbicides. Treatment system effluents will also impact on surface waters.

One of the most serious potential surface water impacts may be siltation and turbidity caused by soil erosion. The critical operational phase is during site preparation and facility construction. Large land areas will have to be prepared for planting each year. Service roads will be installed over the first several years of the farm lifetime. While the erosion potential during this interval is high, much of the damage can be avoided through proper implementation of erosion control techniques. Roads that are sited properly, ditched, and water-barred should minimize erosion hazards. Light-duty roads requiring minor cut-and-fill operations can be covered with wood chips to help stabilize slopes, slow the runoff rate, and reduce the erosive force of rainfall. Since the roads are to be used heavily only at the time of harvest, vegetative cover can be encouraged immediately after planting has been accomplished. The more steeply sloping, northern sites will be especially vulnerable to erosion problems.

Open, prepared fields will also have high erosion potential prior to and during a portion of the first year following planting. Contour ditching to intercept waterflow down long slopes is effective in reducing gullying. All stream courses, lakes, and gullies can be ringed with undisturbed vegetation to reduce silt transport into waterbodies.

Once an area has been planted, ground cover should be reestablished within a reasonable time frame. Irrigation and fertilization are planned to increase biomass yields, and cultivation to control weeds between rows will be necessary during the first year. Cultivation will also increase the erosion potential.

Harvesting and regrowth should cause only minor soil erosion. High-flotation tires on harvesters and wagons can minimize soil disruption.

Work roads may become rutted and bare, however. At some sites they are also likely to be frozen during a part of the harvesting season. In other parts of the country harvesting may be during portions of the wet season. Erosion problems will be greater on these sites, but still much less severe than during the site preparation period.

Runoff volume may increase due to temporary removal of vegetation during land clearing. The removal of leaves, underbrush, and grasses will increase the amount of runoff and decrease the time it takes to reach streams. Since the runoff and the stream velocity will be greater, larger amounts of sediments can be suspended and the erosion potential in the streambed will increase. These problems may only be temporary and probably will be reduced once vegetation returns and the biomass crop begins to grow.

Runoff temperature will also increase during the site preparation phase. Due to the darkness of the disturbed soil and its exposure to the sun, it absorbs more heat and thereby warms the rainfall. Aquatic populations may be affected if stream temperatures are increased significantly.

Fertilizer and soil nutrients may cause enrichment of surface waters. This could occur either from excessive application rates and placement or via runoff. Increased soil nutrient flow will occur mainly during site preparation. Some fertilizer enrichment may occur during the production phase.

Surface water resources will be further impacted by treated effluent from the conversion facility. Exact quantities are difficult to project. There are a variety of chemical treatments which will remove many non-biodegradable compounds. Primary and secondary wastewater treatment systems will treat biodegradable wastes to comply with federal and state standards. The net surface water impact may still, however, be significant, and further analysis of treatment problems are certainly necessary.

Groundwater

Groundwater resources may be affected by various farm management activities. Water tables may fall slightly, and fertilizer and pesticide contamination of groundwater is possible. The greatest potential source of irrigation water for biomass farms is high-capacity wells. Pumping operations may temporarily depress the local water table, even though approximately half of the irrigation water reenters the water table. Lowering of the groundwater level may reduce surface water flows slightly in some areas but probably not enough to disrupt downstream users' riparian rights.

Other Ecological Effects

Various impacts on the ecology may occur either from air and/or water or directly from silvicultural activities. Adverse ecological consequences may include soil and aquatic biota shifts, changes in wetland areas, disruptions of wildlife distribution and habitat, and vegetational community disturbances.

The soil biota might be shifted because of clearing of vegetation, exposure to sunlight, increases in soil temperature, and the application of chemicals.

Aquatic biota could be affected by water temperature changes, increased nutrient and volume flows, and by siltation. The overall effect might be severe enough to cause fishkills, changes in dominant species, and algal blooms if no control measures are implemented. Depending on the existing water quality and the economic importance of downstream aquatic resources, various control measures can be taken to mitigate such adverse ecological consequences. Some level of impact is, however, unavoidable.

Wetland areas may be affected through water table changes and increased runoff rates and volumes, thereby threatening many wildlife species which visit wetland areas regularly. With proper precautions, wetland area changes can be minimized.

Wildlife in the biomass farm area will be impacted both from habitat disruption during clearing and from simple disturbance due to increased noise and human activity. Larger mammals, reptiles, and birds may move out of the disturbed areas to surrounding land. There may be minor losses in small mammal, reptile, and amphibian populations. Some species will benefit from the land clearing; deer populations may increase due to the increased browse supplies.

Wetlands are ecological areas marked by characteristics that are entirely different from other site types. As the name implies, water usually permeates or covers much of the land during major portions of the year. Wetland areas provide habitats for numerous plants and animals, as well as food and water for various animal species. A biomass farm may require such land to be dewatered to allow harvesters and other wheeled machinery to operate. The draining of wetlands could cause irreversible damage to the ecosystem.

Surface water impacts are unique to wetlands, since there may not be any drainage other than that which has been constructed by man. Wetlands are relatively flat. Peat and muck soils impede water flow, and continual siltation in drainage channels may occur. Since runoff would be controlled by ditches, it is likely that some fertilizer and treatment system effluent would also enter the drainage system. The system may connect directly to the ocean or it may feed into slow-moving streams.

As a worst case, the soil-water regime could be severely altered as well as the soil biota. The aquatic biota that inhabit portions of the wetland might be destroyed or displaced to drainage ditches, or to the surrounding, unaffected wetlands. The original wildlife habitat would be eliminated. The typical wetlands vegetation might also be eliminated due to the change in moisture conditions and substitution of the biomass crop.

SOCIOECONOMIC CONSIDERATIONS

A project involving the use of 50,000 to 60,000 acres of land will certainly have some impacts on the region in which it is located. Impacts will result from changes in the use and ownership of numerous moderately sized parcels of land. The change from dairying, row cropping, or forestry to biomass farming will influence the region's economic base, employment, and tax structure as well as recreation demand and basic service needs.

Shifting from cropping and timber harvesting to energy production will influence how the job force in the area is utilized. It will also affect the activity of the service sector of the local economy. This impact will vary depending on the actual economic structure at the biomass farm site.

Population impacts occur when employment opportunities are created, modified, or eliminated. When land that is being used for one purpose is purchased and used for another, major perturbations can be produced. For example, if semiskilled jobs are eliminated and skilled job positions created, the disturbance is great.

From a regional standpoint, the best kind of agriculture is the one that stimulates the local economy by purchasing from local businesses and also selling some of its produce locally. Obviously, small local farms are more likely to engage in local buying and selling than is a biomass farm operator. Much of the machinery and supplies used on the biomass farm would be purchased from a manufacturer or distributor outside the local or regional market. In spite of the enormous capital expenditures necessary to set up and operate a biomass farm, only a small portion of the purchases other than of land will benefit the local economy.

The local economy will be further impacted by the land acquisitions required to initiate biomass production. Upswings in the market value of land probably will occur when sellers or renters realize that a major new force has begun acquiring fee simple titles or lease rights. This upswing will depend on the density of acquisition that is being attempted. If 1 of every 10 hectares is purchased or leased, the average price per hectare is likely to be lower than if 1 of every 4 hectares is purchased (82). Although the sharp upswing effect may thus be mitigated, in the long run land values are likely to increase greatly in the area.

Effects other than an increase in market value may also occur. For example, the local economy should be stimulated to some degree through the influx of lease or purchase funds from outside of the region. Many parcels of land that were being held by local owners may not have been earning rents sufficient to pay the taxes. Both renters and sellers will have a new source of cash which could be traded for material goods at the local level.

The counties in which the biomass farms are situated may experience a significant increase in tax revenue and only a minimal increase in service expenditures. The biomass farmland, whether leased or purchased, should return similar revenues as when it was farmland. A new source of tax revenue will arise from the conversion facility and, since industrial tax rates are usually relatively high, the capital-intensive conversion facility should provide a major new source of county revenue.

A biomass farm would aggregate land parcels under a single ownership. Forest, pasture, and cropland would become uniform in use and management. The uniformity would be the greatest if the entire farm acreage were concentrated in one area to the exclusion of all other uses.

The use of land for biomass production is compatible with other land uses during all parts of the biomass production schedule. Farming, forestry, grazing, and recreation should not experience interference from a biomass land use. If the need for biomass were to diminish at some time in the future, the land could be returned to its previous uses with no decrease in its suitability for those uses.

Social impacts resulting from the introduction of biomass farms into an area differ greatly from the impacts characteristic of many energy technologies. As biomass production more closely resembles farming than it resembles an oil well, public resistance to and adverse social impacts from this activity should be significantly reduced.

During both construction and operation, a majority of the labor force can be locally supplied. Some skilled labor might have to be imported; however, local hiring will help to create favorable community relations between the biomass farm operator and the surrounding residents. Many of those who will be hired may have been laborers or managers of the lands that were acquired for biomass production. These laborers may experience a change in work patterns due to the round-the-clock schedule of some biomass operations. Some of the new jobs may require professional skills that can only be satisfied by importing people from outside the area. Professionals may have cultural and recreational demands different from the local communities.

The most serious social impact may be the increased truck traffic on portions of the rural road system. Noise impacts can be localized and intermittent except along major trucking arteries serving the biomass farm. Throughout the year, large long-haul tractor-trailer trucks may be using roads 16 or more hours per day. This could raise objections from residents who do not benefit from any jobs or land rent payments yet are subjected to the increased traffic and associated noise, vehicle emissions, and dust.

Biomass farms can be fashioned to satisfy recreational and aesthetic demands. Hunting is an important recreational activity in many rural areas. The increased mileage of light-duty work roads will provide better access than was available previously. Since harvesting may increase the food supplies for deer, hunting opportunities may be improved. Other upland game species may also be encouraged. As for aesthetics, the many

roads that will have to be constructed may have a negative effect from a
scenic point of view. This effect should be minimized by the road and
shoulder seeding program which could be carried out to reduce soil erosion.
Ground cover and grasses should be encouraged on the less traveled roads.

4

Overview and Alternate
Sources of Biomass

SILVICULTURAL AND AGRICULTURAL SOURCES

Agricultural crops and wastes and forests were identified as major sources
of biomass in Chapter 3. Managed stands of trees and agricultural crops
are able to convert roughly 0.5–1% of the solar energy covering a given
area into organic matter (99). Cultivated plants are capable of achieving
higher productivities than natural strains of vegetation. Productivities
vary widely with plant type, soil, climatic conditions, and other factors.

Product yield per acre is the principal factor that establishes residue
variation from site to site. Crops produce dry yields in the range of 4–50
tons/acre per year, averaging around 10 tons/acre per year. Crop yields
can be increased through research in plant genetics and improved agricultural and silvicultural practices. On a nationwide basis, crop residues
are principally used as soil conditioners. Almost three-quarters of the
crop residues generated annually in the United States are returned to the
soil. Roughly 20% of the total crop residue yield is fed to livestock without
sale and, of this, a fraction of 1% is disposed of at cost (100).

In forestry, approximately one-third consists of uncollected logging
residues and the remainder in the form of mill residues. Nearly one-half
of the mill residues is sold for various uses, and roughly one-fourth is
utilized as fuel without sale; the remainder is unused.

The potential availability of land for terrestrial biomass production for
energy usage will heavily rely on the competition for other uses for the
land (e.g., grazing and forest products). At present agricultural and forest
products sell for the equivalent of two to ten times their value as energy
products (101).

Table 4.1 lists selected plant species that are considered suitable for
energy production on the basis that they have sufficiently high biomass
yields and/or have little demand as a food source (102). Forage crops

TABLE 4.1 Selected Plant Species of Complexes Having
Potential for Biomass Production

Species	Location
Annuals	
Sunflower × Jerusalem artichoke	Russia
Sunflower hybrids (seeds only)	California
Exotic forage sorghum	Puerto Rico
Forage sorghum (irrigated)	Nex Mexico
Forage sorghum (irrigated)	Kansas
Sweet sorghum	Mississippi
Exotic corn (137-day season)	North Carolina
Silage corn	Georgia
Hybrid corn	Mississippi
Kenaf	Florida, Georgia
Perennials	
Water hyacinth	Florida
Sugarcane	Mississippi
Sugarcane (state average)	Florida
Sugarcane (best case)	Texas (south)
Sugarcane (10-year average)	Hawaii
Sugarcane (5-year average)	Louisiana
Sugarcane (5-year average)	Puerto Rico
Sugarcane (6-year average)	Philippines
Sugarcane (experimental)	California
Sugarcane (experimental)	California
Sudangrass	California
Alfalfa (surface irrigated)	New Mexico
Alfalfa	New Mexico
Bamboo	Southeast Asia
Bamboo (4-year stand)	Alabama
<u>Abies saccharinensis</u> (dominant species) and other species	Japan
<u>Cinnamomum camphora</u> (dominant species) and other species	Japan
<u>Fagus sylvatica</u>	Switzerland
<u>Larix decidua</u>	Switzerland
<u>Picea abies</u> (dominant species) and other species	Japan
<u>Picea omorika</u> (dominant species) and other species	Japan

(Table 4.1 continued)

Species	Location
<u>Picea densiflora</u> (dominant species) and other species	Japan
<u>Castanopsis japonica</u> (dominant species) and other species	Japan
<u>Betula maximowicziniana</u> (dominant species) and other species	Japan
<u>Populus davidiana</u> (dominant species) and other species	Japan
Hybrid poplar (short-rotation)	
Seedling crop (1 year old)	Pennsylvania
Stubble crop (1 year old)	Pennsylvania
Stubble crop (2 years old)	Pennsylvania
Stubble crop (3 years old)	Pennsylvania
American sycamore (short-rotation)	
Seedlings (2 years old)	Georgia
Seedlings (2 years old)	Georgia
Coppice crop (2 years old)	Georgia
Black cottonwood (2 years old)	Washington
Red alder (1-14 years old)	Washington
Eastern cottonwood (8 years old)	United States (eastern and central region), Europe
<u>Eucalyptus</u> sp.	California, Spain, India, Ethiopia, Kenya, South Africa, Portuga, Australia

Source: From Ref. 102.

are utilized as livestock feed; however, a large portion of the land devoted to forage is not suitable for other crops. Forage crop production makes use of the entire aboveground biomass. Other crop production is aimed toward maximum yields of fruit or seed. The majority of forage crops are perennials, which are generally sturdy plant species capable of withstanding unfavorable conditions such as drought, high temperatures, and freezing.

The concept of using agricultural residues as energy sources has already been accepted as a viable approach. California, for example, may become one of the first states to use agricultural waste products as fuel for heating and cooling office buildings (103). The Energy Commission of that state is conducting experiments on a test gasifier for production of low-cost electricity which at the same time serves to reduce air pollution emissions. The system generates gas from peach pits, walnut and almond shells, corn cobs, rice hulls, straw, bark, wood chips, etc.

In the gasification of fuel, combustion releases gases. The hot product gases can then be trapped, cleaned, cooled, and stored for fuel use elsewhere. The product of gasification is a low Btu gas; however, it is a relatively clean burning fuel with a variety of uses. In California alone, more than 27 million tons of agricultural and wood wastes are generated each year. Rough estimates show that this amount of biomass is sufficient to supply nearly three 1000-MW power plants at 30% conversion efficiency (103).

The idea of cultivating woody plants as energy sources on large scale silvicultural energy farms is not a new concept. During the nineteenth century, wood provided the major source of energy in this country. Wood energy crops are not in direct competition for agricultural lands, as they may be grown on marginal lands. They do, however, compete for land which is or may be employed by the present forest industry. This may be a very keen competition, as the world demand for paper and paperboard has been estimated as likely to exceed 400 million metric tons (tonnes)/year within a quarter of a century (102)—almost double the amount used presently. There will be increasing demand not only for pulp but for lumber as well Since longer growing cycles would be required for producing wood of the quality necessary for the various nonenergy uses, considerable competition for land suitable for wood production could exist. It is conceivable that energy farms may not be able to compete within the forest industry because of the higher prices which these nonenergy uses may realize.

Because of rather recent advances in silviculture, wood biomass yields per acre per year have been dramatically increased. As an example, short-rotation, densely grown hardwoods have been generated at yields of some 10-16 tons/acre per year at costs per unit energy that are comparable with coal in the Pacific Northwest and the Southeast (105). Such yields strongly suggest that competitive production can be achieved. The yields produced were obtained without genetic selection from wild seedling populations. No irrigation was used, and only a minimum of weed control and fertilization was employed. The wood quality generated is not suitable for use as lumber and meets only marginal quality standards for pulp and particle board feedstock. Within this framework, woody plants for energy may be capable of competing with conventional energy uses.

Methods of Increasing Crop Yield

As noted earlier, the photosynthetic process is relatively inefficient in comparison with direct solar energy conversion systems which can operate in the range of 40-50% efficiency. Biomass does, however, provide a storable, readily transportable form of solar energy.

Carbon dioxide availability has been observed to be a limiting parameter in the photosynthetic process efficiency. To improve crop yields, various methods have been proposed for supplying CO_2 to plants. It should be noted that the upper theoretical efficiency limit for the conversion of solar radiation to biomass is on the order of 8% of the total amount reaching plants based on C_4-chain structures and 5-6% of that reaching C_3-type plants (105). C_4-type plants photorespire at a slower rate than C_3-type plants do, presumably because of a cellular mechanism in the former which deters C_2 from escaping their tissues. Crops falling under the C_4 category include corn, sugarcane, and sorghum, which are primarily limited to tropical and semitropical regions. Corn is probably the most widely used C_4 plant and is generally a hardier species than other C_4 crops. The majority of C_4 plants are capable of converting only 2% of the isolation that reaches them. The C_3 crops on the average convert only about 1% of the total available light (105).

Application of CO_2 in greenhouses was attempted nearly 25 years ago on tomato, lettuce, and cucumber plants with increased yields of 25-100%. Application of dry ice, microcapsules, or mixtures of CO_2 with water in trickle-irrigation systems have been proposed for increasing various outdoor crop yields.

Other attempts at increasing crop yields have involved the genetic rearrangement of leaf display. Attempts have also been made to modify and optimize metabolic processes through biological and genetic changes.

Hybrid development is an area of considerable interest in that it offers the potential of lengthening the growing season of various species. One approach involves controlling the output of glycolic acid from C_3-type plants. Optimistic predictions have been made that the yield of C_3 plants such as wheat and soybean can be increased by more than 50% (101).

Molecular biology is being applied to the selection of plant cells that display lower photorespiration rates. Genotype selection appears to be a promising approach to establishing plant growth from single cells. Limited success has been achieved in regenerating whole plants such as tobacco, sugarcane, southern pine trees, and Douglas fir. Cell fusion between

related and unrelated species offers some promise of generating strains of
hybrids having unique characteristics, some of which may have commercial
value. At present, no plants have been cultivated from fused cells of unrelated species.

Nitrogen fixation is another area of considerable interest. In nitrogen
fixation, certain strains of bacteria interact with certain plants causing
nitrogen to be extracted from air, forming ammonia and nitrates which are
basic plant nutrients. These bacteria exist in nodules or locate near the
roots of nitrogen-fixing plants. Various microbiological techniques are
being tried in an attempt to maximize the development of nitrogen fixers
which conceivably could be transferred to nonnitrogen fixers such as corn.
With the proper chemical catalyst, the action of the bacterial enzyme nitrogenase might be initiated, thereby establishing such nodule growth on currently non-nitrogen-fixing plants. If successful, this would greatly reduce
the dependence of agriculture on petroleum products for nitrogen-based
fertilizers. At present, agricultural industry is probably one of the largest
consumers of petroleum in the United States, the reasons being its high
level of mechanization and the widespread use of petrochemical-based fertilizers.

Fertilization and Silvicultural Practices

Fertilization has long been recognized as a means of enhancing the biomass productivity of plants. For short-rotation forest management it is
also a means of restoring the original fertility of a site which has undergone
several short rotations.

White et al. (106) have noted that fertilization requirements for hardwood
species employed in short-rotation systems are not well outlined. It is
known, however, that most species will favorably respond to fertilizers.
The extent of response will depend on existing site conditions and the specific
nutrients used.

Sycamore seedlings planted on a poor site have been shown to respond
dramatically to nitrogen fertilization. Kitzmiller (107) conducted a study
in which seedlings were planted in 4- by 4-ft areas on a site in the North
Carolina Piedmont. The particular site chosen was one of low fertility.
Nitrogen was applied at annual rates of 100 and 200 lb/acre. Stem volumes
were measured at the termination of each growing season. Stem volumes
of nitrogen-fertilized trees were found to be 650 and 1570% greater for the
100 and 200 lb/acre per year application rates, respectively. Several
sources of nitrogen were studied by Kitzmiller, including urea, ammonium
sulfate, ammonium nitrate, and sodium nitrate.

No significant growth response was observed to liming at the rate of
2000 lb/acre in the first year. Significant response to phosphorus fertilization was not observed until the third consecutive year of treatment.

Bonner and Broadfoot (108) tested nitrogen, phosphorus, and potash
applications on eastern cottonwood seedlings in sand culture. Maximum

response was found with applications of 200, 150, and 200 lb/acre per year of nitrogen, phosphorus, and potash, respectively.

Response of eastern cottonwood seedlings to liming on a coastal plain soil was also tested. After a period of 2 years, trees grown in limed pots showed only about 17% height increases over controls. The investigators attributed poor response to the time and method of application. In the study limestone was spread between rows of trees and dished lightly, resulting in shallow incorporation. Standard agricultural practice is to apply lime prior to planting to produce deep incorporation. Calcium is relatively immobile and as such requires mixing with the soil in order to benefit crops.

Heilman et al. (109) found black cottonwood to show significant response to fertilization prior to planting. Three treatments were employed in this study: (1) a control run with no fertilizer application; (2) application of 431 lb/acre of nitrogen derived from urea; and (3) application of 431 lb/acre of nitrogen, phosphorus, and potash in a commercial fertilizer. Trees were harvested at the end of a 2-3 year period. The total aboveground biomass, excluding foliage, was found to be 20% greater than the control for trees treated with nitrogen fertilization and 33% greater for those treated with the commercial fertilizer mixture.

Eucalyptus plantations have been experimented with in the United States. Rock phosphate, diammonium phosphate, and 10-20-10 fertilizer have been employed in such experiments on flatwood sites (110, 111).

Loblolly pine has shown good responses when nitrogen and nitrogen/ phosphorus were applied to newly planted stands on lower coastal plains soils (112, 113). The application of nitrogen alone can stimulate competitive herbaceous vegetation (114).

In general, little is known about fertilization criteria and potentials to short-rotation biomass generation. Requirements and opportunities are likely to be site specific. Further research is warranted on growth responses to fertilization for most plant species in order to permit adequate evaluation of the results of fertilization in any specific situation and site.

CULTURAL PRACTICES OF
SILVICULTURAL/AGRICULTURAL BIOMASS

A variety of parameters have to be controlled during the life of a crop or stand in order to ensure optimum growing conditions, increase yields, and protect plantlife from damaging conditions. Some of these measures include fertilization, competition control, irrigation, pest protection, disease, and fire.

Most sites chosen for present forest and agricultural sites tend to be inherently fertile. In general, competitive vegetation as well as the marketable growth will flourish on such sites. Hence, for most biomass farms, the control of competitive vegetation will be an integral part of the overall plantation operation.

The control of competition for such variables as sunlight, soil moisture, and nutrient supply can have significant effects on mortality and crop yields. Such control starts with site preparation but continues after planting. There are three general approaches, namely, mechanical control, chemical control, and crop competition techniques.

Mechanical control is basically the cutting and/or burying of weeds. During the first year of rotation on a short-rotation farm, conventional tractor-drawn agricultural equipment (e.g., disks, harrows, plows, cultivators, and rotary hoes) can be used. After the first growing season, mechanical control may be difficult to maintain on silvicultural farms without major modifications to equipment. When the second growing season begins, crop trees might be on the order of 3-5 ft tall, and the close spacing necessary for short-rotation biomass production may make it difficult to run equipment between rows.

Chemical control involves the use of herbicides which are capable of reducing or eliminating competing vegetation. When herbicides are selected and applied properly, they do minimal damage to crops. Chemical control can be employed in cases where the spacing and size of crops preclude the use of mechanical alternatives. They also provide an alternative for use on wet sites, where cultivation may be difficult and heavy equipment can cause severe damage through soil compaction. Herbicides vary in their toxicity to different weed and crop species.

Preemergence herbicides are chemicals which must be applied to the bare soil. The degree of herbicide application varies with the specific chemical selected and the site conditions. These types of herbicides will not affect weeds which have already sprouted.

Contact herbicides must be used on those weeds which are visible. These chemicals are generally applied as a directed spray. Preemergence herbicides can be applied either by aerial methods or on the ground using a tractor or hand spray. Contact herbicides must be carefully directed at weeds and away from the crop.

A large number of weed species are unable to survive in shaded environments. In the case of silviculture, once a short-rotation tree crop has become fully established and crown closure occurs, crop competition control becomes less demanding.

Heiligmann (115) has outlined major parameters and factors of a weed control program. These are illustrated in Fig. 4.1.

Crop-Damaging Agents

The protection of biomass farms from various damaging agents such as wildfire, insects, and diseases is a major consideration. The dense, widespread, and continuous monoculture that would characterize such energy plantations might present certain risks in terms of exposure to destructive conditions.

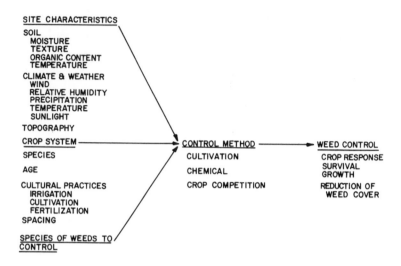

FIG. 4.1 Major parameters and considerations in establishing a weed
control program.

In the case of silviculture biomass farms, the understory vegetation
presents a threat from fire. Crop trees themselves are unlikely candidates
for fire because of their higher moisture content, except perhaps during
periods of drought. During the early fall, herbaceous understory can dry
out and such material can readily burn, causing extensive damage to crop
trees.

In farms using wide spacings, ground cover may present a large fire
hazard. In this case, the installation of firebreaks may be necessary.
Frequent patrols by plantation personnel during fire seasons may be required. The risk of fires occurring on a short-rotation farm should be low;
however, such areas should not be used for recreational purposes that would
expose crops to the risk of man-caused fires.

Fire suppression can be carried out using conventional methods. For
example, backpack-carried sprayers and truck-mounted tanks might be
used. Aerial suppression with water or chemicals would probably be appropriate for certain situations.

Other damaging agents include various types of insect infestations.
Various treatments or manipulations can be used as preventive or control
tools to curtail insects before they become destructive.

Insects which attack trees are of three main types:

1. Defoliators
2. Sapsuckers
3. Shoot-stem borers

Defoliators are among the most common tree-attacking insects. Included in this group are the larvae of moths and sawflies as well as the larvae and adults of beetles. Defoliators are well adapted to feeding on trees, since they can convert ingested food to metabolites on their preferred hosts. This means that they have a high survival rate and populations tend to increase after initial attack. They can damage the foliage in several ways. First, they can devour the leaf or chew holes in it. Other types eat only on the surface, or between the veins or inside. Some defoliators may reduce the total leaf area and, therefore, the photosynthetic capacity of the tree which impairs growth.

Sapsuckers include aphids, scales, spittlebugs, and mites. They are adapted to tap sources of metabolites in the phloem of the tree. They remove sugars and amino acids, and there are substances in their saliva (e.g., phenolic compounds, hormones, and enzymes) which depress the metabolism of the tree, causing reduced leaf size and reduced wood formation.

Shoot-stem boring beetles and moths are usually low-density pests that localize in specific plant tissues. They are capable of killing trees by selective feeding habits.

Control of insect pests in short-rotation crops should begin with proper site preparation. Effective site preparation techniques can increase the vigor of plants, rendering them more resistant to insect attack. Fertilization can also have a highly variable impact on insect attacks. In general, defoliator population development is limited in fertilized stands. Insect control would generally be accomplished through the use of insecticides. A variety of insecticides are commercially available. The choice of an insecticide, its formulation, and mode of application will depend on its toxicity for the insect species in question, insecticidal selectivity, hazard to man and other nontarget organisms, phytotoxicity, residual persistence, cost, convenience of use, and speed of action. Such a choice will be specific to a given situation and should be made with the aid of a competent entomologist.

Insecticides can be applied either in liquid form or as powders. Application may be performed by terrestrial means with sprayers or power dusters, or by aerial means. Aerial applications are most appropriate for combatting defoliators, in which case either concentrated liquid or dusts may be used. Most of an insecticide applied by aerial means is deposited on foliage and branches. Insecticides may also be applied in liquid or powder form directly to the soil to combat soil-borne insects.

Many diseases, particularly of forest trees, are caused by fungi. Varieties of fungi cause decay and/or defoliation. Canker-causing fungi can kill a tree either directly by girdling or by so weakening them at the point of infection that breakage occurs. Bacterial and viral diseases of forest trees are known to occur but are far less common than fungal diseases. Viral diseases may present some of the greatest threats to other kinds of biomass (nonsilvicultural) farms.

Proper site preparation as well as maintenance of proper soil moisture
and nutrient levels should prevent most disease outbreaks. Treatment with
commercial fungicides, applied on the ground or from the air depending on
the type of disease, may be used should an outbreak occur. In forest biomass, the removal of infected trees may be necessary.

Animals can also be included in the category of damaging agents. Small
animals can damage young seedlings or sprouts by feeding on them, while
larger animals can trample them. Deer and cattle might be deterred by
erecting fences or by using brush fences. These can be constructed from
material which would otherwise be burned in land clearing or slash treatment.

Damage by smaller animals (e.g., rabbits) would be confined to newly
planted seedlings or sprouts. Smaller animals might not pose a major
problem to short-rotation plantations.

AQUATIC BIOMASS

Conventional crops are limited in biomass production due to the fact that
nutrients must be absorbed from the air and soil. Algae, on the other hand,
have a greater ratio of nutrient-absorbing surface area to plant volume and
so are not as limited in absorbing nutrients and sunlight. Algae are therefore characterized by rapid growth and reproduction. Oswald (116) has
reported that yields of 70 dry tons/acre per year are attainable. As noted
earlier, algal yields are not only promising in terms of biomass supplies
but also as a source of protein. On an acreage basis, the annual algal protein yield may be roughly 127 times greater than that of soybean, 540 times
greater than that of wheat, and 560 times greater than that of rice (117).
Algae-derived protein has been determined to be digestible and as efficient
as other more commonly used proteins.

Algae is also a useful material in sewage treatment facilities, specifically
as an oxidizing agent in stabilization ponds. In the process of aerobic decomposition of sewage wastes, bacteria use oxygen to break down sewage
organisms. By-products of the reaction include ammonia and CO_2. The
algae use nutrients to produce oxygen, which is in turn used by the bacteria
for reproduction.

Conventional stabilization ponds typically are at least 5 ft deep and
detention times are on the order of 20 days. However, in order to maximize
algal growth in stabilization ponds, shallower depths are necessary to permit sunlight to penetrate throughout; in addition, shorter detention times
are necessary to maintain high nutrient levels (118). A theoretical algal
culture system was briefly discussed in Chapter 3.

For many years, algae and photosynthetic bacteria have been recognized
light-driven hydrogen-producing capability. The mechanism of

hydrogen evolution of algae is directly linked to the process of photosynthesis. There are several distinct advantages to using marine photosynthetic biomass as sources of fuels, the most apparent being the abundance of salt water. Many areas of the world suffer from deficiency of fresh water, and so the ability to use salt water becomes a significant consideration. Salt water contains a variety of nutrients, including CO_2, magnesium sulfate, and potassium, essential materials to the growth of photosynthetic microorganisms.

Fuel produced from marine microorganisms is essentially energy at a low cost. In addition, the conversion methods described in the next chapter do not require the use of fossil fuels.

Photosynthetic bioconversion processes are linked with the production of a variety of valuable by-products. Some of these by-products offer an opportunity for a profitable multiutilization plan (i.e., chemicals for medicine, food, and fuel). Algal carbohydrates can be converted to alcohol and methane gas. Most organic materials can be transformed into a mixture of CO_2 and methane by anaerobic bacterial fermentation. The process involves specific action of at least two types of bacteria—the same principle as has already been applied to organic wastes.

Tropical and subtropical regions have produced marine microorganisms with unique characteristics. Marine waters in tropical regions contain a great diversity of photosynthetic microorganisms. These areas have a continual supply of high-intensity solar radiation. Weather conditions allow convenient year-round collection of biomass.

The maximum theoretical synthetic solar conversion efficiency is on the order of 30% (119). The problem of increasing bioconversion efficiency involves a wide range of variables including growth rates and rate-limiting reactions.

Growth rates of plants are related to their photosynthetic capability. This, in turn, determines a plant's hydrogen-producing and nitrogen-fixing potential. Obviously different plant species exhibit different growth potentials. Even among the same species of plants, potentials may vary significantly depending on environmental conditions such as sunlight, temperature, salinity, and nutrient availability. Plants from different regions may exhibit different environmental optima. As an example, northern plant species may require less solar energy intensity to reach the same photosynthetic rate as compared to southern species.

High growth potentials and yields have been reported for freshwater vascular plants. One example is the water hyacinth. Lecuyer (120) has reported yields of 60 tons/acre per year in warm climates. Such plants have also been found to be useful in controlling eutrophication in natural bodies of water through nutrient removal.

Overview and Alternate Sources of Biomass

TABLE 4.2 Manure Composition and Daily Production

Type Animal	Daily Production (per 1000 lb live animal) Wet Weight (lb/day)	Volatile Solids (% wet weight)	Nitrogen Solids (% wet weight)	Phosphorus Content (% wet weight)
Dairy cattle	76.9	7.98	0.38	0.10
Beef cattle	83.3	9.33	0.70	0.20
Swine	56.7	7.02	0.83	0.47
Sheep	40.0	21.50	1.00	0.30
Poultry	62.5	16.80	1.20	1.20
Horses	56.0	14.30	0.86	0.13

Source: From Ref. 122.

FIG. 4.2 Front-end loader filling dump truck with leaves piled by a street sweeper.

FIG. 4.3 A front-end loader dumps compost material from windrow directly into a shredder-mixer.

PAPER WASTES

It is possible that wastepaper, a source of cellulose, might serve as a raw-material feed source for anaerobic digesters in rural areas of developing countries. The digestibility of paper is relatively high. Manufacturing processes normally remove the majority of lignin present in the initial wood. Wastepaper might be a significant source of carbon in the anaerobic digestion process or might serve as a supplement fuel with animal wastes.

MATERIALS OF ANIMAL AND HUMAN ORIGIN

Animal wastes generated in the agricultural industry can be a significant source of substrate for bioconversion processes. Manure is already the major feedstock for many digesters in developing countries.

Animal wastes represent a source of carbon and a potential source of nitrogen which is necessary for the operation of an anaerobic fermentation process. Various properties, production quantities, and nitrogen and phosphorous contents for different animal wastes are given in Table 4.2. As noted in Table 4.2, both composition and quantity vary with animal species. In addition, animals of the same species under one type of controlled growth or diet will produce manure and urine markedly different in content and quantity than those grown under different conditions (121).

Night soil (human feces and urine) constitutes still another raw material source that may find use in anaerobic fermentation operations for the production of methane. Night soil is similar to animal manure in nitrogen and phosphorus content.

Other sources of urban-related biomass are shown in Figs. 4.2 and 4.3.

5

Direct Combustion of Biomass

OVERVIEW OF CONVERSION TECHNOLOGY

Conversion processes can be either physical or chemical or a combination of both (refer to Fig. 5.1). Physical conversion techniques are aimed at physically altering the biomass form. For example, the form of silvicultural biomass or wood can be changed by: (a) size-reduction techniques such as chipping or pulverizing; (b) drying to reduce water content; (c) screening or selecting specific components or species of the biomass for certain properties; or (d) combinations of these approaches. The end result of such physical processes is a prepared biomass suitable for combustion.

Chemical conversion techniques are aimed at altering the molecular structure of the biomass. Examples include combustion operations to produce thermal energy; incomplete combustion to produce chemically relative products [synthesis gas (see Chap. 8)]; fragmentation of cellulose molecules in an aqueous slurry through the action of mineral acids, selected bacteria, or selected enzymes.

Energy-related products that can be produced by physical and/or chemical conversion methods include electricity, fuel gases with varying heating values, and various liquid and solid fuels. These energy-related products are strong alternative candidates to replace imported petroleum and domestic natural gas. Ammonia and methanol, for example, can be derived from wood biomass. Although they are not normally considered fuels, they are likely substitutes for natural gas. Ethanol, another wood-based chemical, might also decrease the demand for petrochemical feedstocks such as ethylene. The various fuels and conversion technologies for biomass are discussed in Chapter 7. This chapter is specifically concerned with techniques that directly convert biomass to energy.

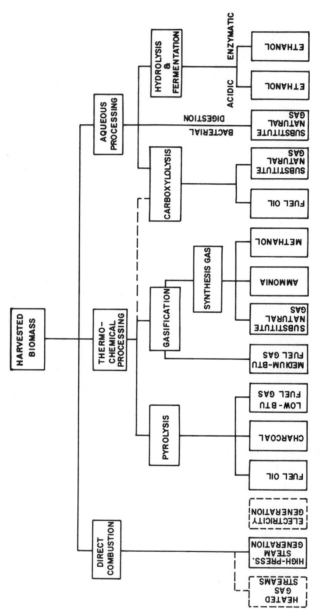

FIG. 5.1 Various conversion approaches as applied to wood biomass. (From Ref. 123.)

CONVERSION OF SILVICULTURAL BIOMASS

Combustion is the traditional approach to converting wood biomass to energy. The basic end product of combustion is thermal energy, usually in the form of steam generated at elevated pressures and temperatures from the heat of combustion of the wood. The steam generated is suitable for various purposes such as heating or for the generation of electricity. Hot combustion gases generated are suitable for heating or drying purposes.

The direct conversion of wood biomass to energy via open fires for the generation of heat is an ancient concept. At present, wood combustion is carried out on a large scale, producing steam that has value on an industrial level. The generation of steam can be accomplished by two approaches: (1) through the use of conventional boiler-furnace generation, in which heat removal is accomplished by radiant heat transfer to absorbing surfaces; and (2) by fluid-bed incineration, which involves the production of heated gases from wood combustion. These gases are transferred to banks of tubular heat-absorbing surfaces which accomplish heat recovery mainly by convection.

High-pressure steam is required for industrial purposes and to provide electricity generation. The generation of electricity from wood biomass in central power stations is almost nonexistent in this country; however, industrial steam generation is widely practiced in both the wood industry and the pulp and paper industry.

Equipment generally used for burning wood can be classified as field-erected boilers, packaged systems, and special combustors (123). Large steam boilers commonly employed in the pulp and paper industry are usually custom-engineered for individual applications and are assembled on site. The capacities of some of these large field-erected boilers designed for firing coal can exceed 10^6 lb/hr of steam. Industrial boilers for firing only wood are limited to about half of this capacity (123).

In general, field-erected boilers for firing wood consist of larger combustion chambers that are considerably more complex and often more costly than oil- or gas-fed boilers. Wood-fired systems are similar to coal-fired units in design. Both systems have low volumetric heat release rates, on the order of 25,000 Btu/(hr · ft^3), and both are designed for complete combustion. Ash removal and thin-pile burning is usually accomplished by vibrating, reciprocating, oscillating, and/or traveling grates. For wood fuels, pinhole grating systems for supplying combustion and cooling air are preferred, as wood fuels have varying and often high moisture content.

The size and moisture content of residue fuels which can be combusted in field-erected boilers depends on a number of factors, the major ones being:

1. Design of the combustion chamber
2. Preheat temperature of the air
3. Combustion properties of the biomass

Moisture contents of biomass in excess of 60% are undesirable, as they can cause blackout conditions or smelt water explosions. Most units, if properly designed, can handle residues with moisture content as high as 50-55%.

A typical design of a wood/bark-fueled or bagasse-fueled water-tube steam generator widely used in the pulp and paper and sugar refining industries is illustrated in Fig. 5.2. The design shown closely resembles coal-

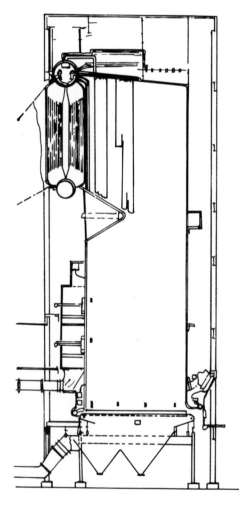

FIG. 5.2 Typical wood/bark-fueled water-tube steam generator employed in the pulp and paper industry.

fired units of comparable size in which coal fuel is unpulverized. These systems can include in their designs tubular air heaters for preheating the combustion air. A steam superheater is located ahead of the steam generator convection bank tubes where gases exit the furnace.

Firing techniques vary but are all based on the spreader-stoker concept. Wood fuel is fed to the system by spreading it pneumatically, or mechanically, across the combustion chamber onto the surface of a traveling grate. Smaller particles will tend to burn in suspension, and larger pieces will fall onto the grate. The design of the feed system is critical and should provide an even, thin bed of fuel on the grates. Flames that extend over the grates radiate heat back into the fuel, assisting combustion. Underfired and overfired air are both employed in controlling the combustion process. Furnace walls are lined with heat-exchange tubes, and because the system is designed with little refractory material, the furnace will respond to load variations rapidly.

Three types of grate systems are most often employed. These include the dump grate, stationary grate, and the traveling grate. Dump grates are limited to an upper air preheat temperature of approximately 400° F when firing at about 50% moisture content fuel. They have an advantage over stationary grate systems in that they can be cleaned more readily.

Stationary grate systems consist of an arrangement of water tubes interconnected with the steam generator circuit. This arrangement allows air preheat temperatures to exceed 550° F without damage to the grate. Stationary grates can be horizontal or inclined. The latter arrangement permits the fuel to slide to the discharge point for the ash by gravity.

Traveling grates are the most widely used of the three designs. The system provides for continuous dumping of ash. This provides more effective ash removal and leads to longer equipment life.

The type of fuel that is used in the system impacts on the required grate area. Bliss and Blake (124) report that, with coal, liberation of 8.50×10^5 Btu/hr per unit grate area is achievable. For wood at 55% moisture content and some 2-3 in. in length, the liberation can increase 1.11×10^6 Btu/hr per square foot.

Stoker-firing equipment are limited to a maximum capability of about 600,000 lb/hr for the steam generator (124). Bliss and Blake note that if a double grate were to be employed side by side (a design which would require two feed streams for the wood biomass), capacities could be doubled.

Packaged or shop-fabricated boilers consist of either firetube or water-tube units. These systems are factory built and hauled to the site intact. Physical dimensions are primarily limited by transport clearances; hence, firetube boilers generally do not produce more than 30,000 lb/hr of low-pressure steam. Water-tube boilers, however, are capable of about 100,000 lb/hr or more of high pressure steam. Size restrictions limit the use to single units in applications that require large steam rates, typical of the pulp and paper industry. The biomass generally employed is hog fuel

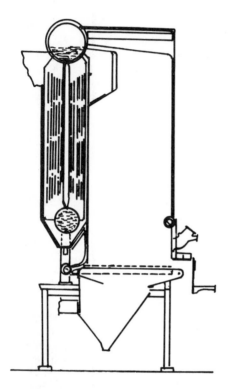

FIG. 5.3   A shop-fabricated vertical furnace design for wood biomass.

with particle sizes around 2-in. or less and moisture contents typically
under 40%. Two types of water-tube designs are illustrated in Figs. 5.3
and 5.4. The vertical design shown in Fig. 5.3 is capable of adapting to
different stoker designs. Horizontal and inclined systems have been constructed for capacities of up to 75,000 lb/hr of steam at 650 psig and 750° F
effluent steam temperature (124).
    A design which has evolved fairly recently is a variation of the vortex
combustor. In these units a forced vortex is generated by the introduction
of compressed air tangentially into the combustion chamber. Fuel (wood
residue or other solid fuels) is fed to the combustion chamber by compressed
air. High combustion efficiencies (on the order of 90%) have been claimed
by manufacturers, who attribute this to the turbulent and detention time of
the fuel within the combustion chamber. Units range in output from 5 to
100 MBtu/hr, and both horizontal and vertical designs are common (123).
Vortex combustors are usually stand-alone units. Their primary components include an air blower, combustion chamber, operating controls, and

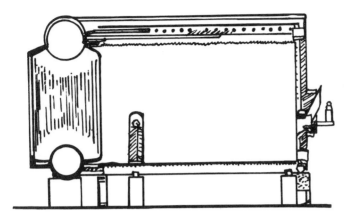

FIG. 5.4 A shop-fabricated horizontal furnace design for wood biomass.

FIG. 5.5 One type of bioconversion system—the Energex Burner System:
(1) the combustion chamber; (2) combustion air inlet and blower assembly;
(3) wood fuel metering bin; (4) rotary air lock feeder; (5) blower to convey
wood to burner; (6) combustion air manifold. (Courtesy of Energex Ltd.)

various auxiliaries. Liners typically employed in the combustion chamber
are of the refractory type. Ash is removed either in solid form or as liquid
slag. Applications include direct kiln or rotary driers in addition to boilers
and heat exchangers. Hogged wood fuel 1 in. in size or less with moisture
content under 40% is the primary wood biomass used.

Another type of bioconversion system is illustrated in Fig. 5.5.

FOULING AND FIRING CONSIDERATIONS
(HOG FUEL BOILERS)

In the lumber industry, roughly 50% of wood biomass is removed from a
log to produce a board. Mills often use sawdust or mixtures of sawdust
and shavings as fuel burned in boilers for steam production. This type of
wood biomass fuel has the advantage that it does not require further physical
preparation before being fired. The majority of wood wastes require further
size reduction in a hog. Size reduction is necessary to facilitate storage,
feeding, combustion, etc. Size-reduced materials are often mixed with
varying amounts of sawdust and shavings, the final mixture constitutes hog
fuel. Table 5.1 shows typical heating values for hogged fuel along with
elemental analysis for different wood species (125).

There are a number of considerations that must be taken into account in
order to burn wood biomass efficiently. The moisture content in wood, as
previously noted, can greatly restrict the combustion efficiency. Harvested
wood biomass generally contains roughly 50% moisture, and the heat necessary to dry wood can be on the order of 1500-2000 Btu/lb of water removed.
This can represent anywhere from 15 to 20% of the heating value of dry
wood alone. In addition, as noted in Table 5.1, wood has a relatively high
oxygen content. This large source of oxygen can lower furnace temperatures, resulting in a redistribution of the heat transferred by radiation in
the furnace and convection in the superheater, economizer, and air heater
sections. This can result in a derating of the boiler capacity. (Note, however, that this can have the favorable effect of reducing nitrogen oxide
formation.)

Ash composition analyses on wood tends to suggest that fouling may be
a moderate problem. Wood ash typically consists of approximately 50-60%
CaO. It is also relatively high in Na_2O and K_2O, both in the range of 4-7%
(126). If wood biomass is substituted for residual fuel oil, increasing fouling problems on heat-receiving surfaces can be expected. Using wood in
boiler furnaces that are designed for pulverized coal is not likely to present
any more difficult fouling problems than already experienced with coal
alone; however, as noted by Hall et al. (126), alkalies may tend to accumulate more readily in regions such as the superheater, where temperatures
are lower.

TABLE 5.1 Heating Values and Elemental Analysis of Various
Hogged Fuels

Item	Western Hemlock	Douglas Fir	Pine (sawdust)
Hydrogen (%)	5.8	6.3	6.3
Carbon (%)	50.4	52.3	51.8
Nitrogen (%)	0.1	0.1	0.1
Oxygen (%)	41.4	40.5	41.3
Sulfur (%)	0.1	—	—
Ash (%)	2.2	0.8	0.5
Heating value—dry (J/kg) 10^{-6}	20.05	21.05	21.24

Source: From Ref. 125.

Ananth et al. (127) point out that there are several options available to
retrofitting existing systems with wood biomass as fuel. Substitution of
wood in existing coal-fired stoker boilers can be done with only minor boiler
modifications being necessary. Probably the most elaborate modification
required is the addition of a wood-feeding scheme or firing parts. It is
also possible to fire wood in suspension in systems designed for pulverized
coal or heavy oil. The wood must undergo size reduction to less than 0.25
in. and requires drying to ensure rapid combustion. Suspension wood-fired
boilers have already been discussed. Firing wood in boilers designed to
be fed with natural gas or light fuel oil is possible; however, extensive
boiler modifications would be necessary, particularly in the area of handling
ash.

FLUID BED COMBUSTION

Figure 5.6 illustrates the general arrangement of a fluid-bed combustion
system for the generation of steam. Combustion and steam generation steps
are separated in this system. In the arrangement shown, combustion generates hot gases, which are transferred to an adjacent boiler. The boiler
collects the heat from the gases and produces steam before discharging
gases to the stack.

Fluid-bed steam generation has been proposed for low-grade wood wastes
generated in the pulp and paper industry for which there are no superior

[]108   Direct Combustion of Biomass

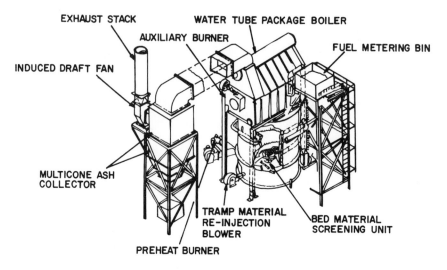

FIG. 5.6 A wood-fired fluid-bed steam generating system.

uses in various manufacturing operations. Such residues may have moisture
contents as high as 65% and contain varying amounts of inert substances.
    These systems are known to display stable operation but at the cost of
careful size preparation and control of gas velocities. Obviously in an actual
mill operation the system must have the capability of handling irregularly
shaped wood biomass and a variety of inert materials of unpredictable char-
acter. To do this, the fluid bed is designed with a prepared mixture such
as sand and the residue, which is fed in at a slow uniform rate. Combustion
air serves as the fluidizing medium. Bed temperatures are typically in the
range of 1200-1800° F. Excess air and moisture content must be carefully
controlled so that bed temperatures do not approach melting or fusion conditions.

COMBINED FIRING SYSTEMS AND RETROFITTING

Combined firing systems involve the use of biomass and other fuels such
as coal or oil in the production of steam or electric power in utility boilers.
Figure 5.7 illustrates the general scheme for combined firing systems.
The concept has been used for many years in the paper industry. In this
case fuel oil is employed as a supplemental fuel in recovery boilers with
concentrated black liquor.
    In small-scale industrial boilers, wood wastes (i.e., wood chips) have
been fed with coal. This approach usually requires no waste processing.

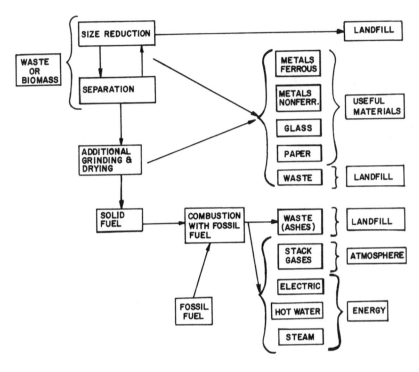

FIG. 5.7 General scheme of combined firing systems.

Systems that use bark and coal require that the bark be prepared by shredding. Wood-coal combined firing systems are generally located in areas where there are adequate supplies of wood wastes.

The pulp and paper industry is unique in that it exhibits marked differences in relationships between its consumption of fossil fuels, electricity, and biomass for fuel. Mills located in the South account for approximately 62% of the energy purchases within the industry. Biomass in the form of process wastes provide approximately 50% of the fuel requirements. Paper mills situated in the North Central region purchase approximately 11% of the total energy employed in U.S. paper manufacturing. Also in this region, coal constitutes roughly 37% of the fuel required (123). Zerbe and Arola (123) note that the North Central region, aside from having a large coal-burning capacity which might be readily converted to burning wood, is also characterized by great diversity of the pulp and paper industry and a large land area.

Inman et al. (128) have examined the options of retrofitting small oil-fired, gas-fired, and coal-fired power plants to burn wood biomass. They note that capital subsidies do not have a major impact on either supply or coal plant retrofitting because of the relatively low proportion of costs attributed to capital in their operation. Capital subsidy operations may however become attractive as biomass supply costs become more capital intensive with the development of new harvesting techniques. Certainly the option is available of replacing or retrofitting small existing oil/gas-fired power plants with facilities using agricultural residues as well.

ELECTRICITY GENERATION FROM WOOD BIOMASS

Federal policy is already aimed at the expanded use of coal for electric generation both in existing and new power plant installations. The potential for wood biomass in electrical generation is high, especially in light of the environmental concerns for SO_x and NO_x emissions and associated high costs of pollution control equipment.

A modern steam-turbine/electricity-generation cycle is illustrated in Fig. 5.8. At present, it is proposed that steam generators in such cycles be coal-fired in the coming years. However, coal containing an excess of 0.6 lb of elemental sulfur per MBtu of heating value will generate sulfur oxides that exceed current stack gas emission standards.

There are a variety of approaches to handling SO_x emissions, including wet scrubbing techniques and electrostatic precipitators; however, removal is not straightforward. The wet limestone process illustrated in Fig. 5.9 is one specific flue gas desulfurization process under consideration.

The basic approach to burning coal in a boiler designed for electrical generation involves feeding the fuel in suspension. The coal must be pulverized with particle size distributions on the order of 65% through 200 mesh, 85% through 100 mesh, and 98% through 50 mesh (125). Coal ash can be removed from the furnace either as a liquid slag or as a powder, depending on furnace temperatures. If it is removed as a liquid slag, the furnace is referred to as a wet bottom unit; if a powder, as a dry bottom. A typical unit based on coal firing may be on the order of 880-MW generating capacity, which would require about 6.4×10^6 lb/hr steam generating capacity at 3515 psia, 1000° F superheat, and 1000° F reheat (127).

It is likely that wood biomass used for electrical generation would be processed in pulverized form, analogous to coal. Ananth et al. (127) points out that this approach would tend to reduce major modifications or premature equipment obsolescence to large-scale use in central power generating stations. However, the feasibility of pulverizing wood biomass has not been fully assessed. Sanding dust and sawdust only approach the size range of pulverized coal. In addition, preparation of both these materials is costly and requires dry wood.

The manufacture of wood flour is based on limited industrial experience (wood flour is used in linoleum manufacturing, explosives formulations,

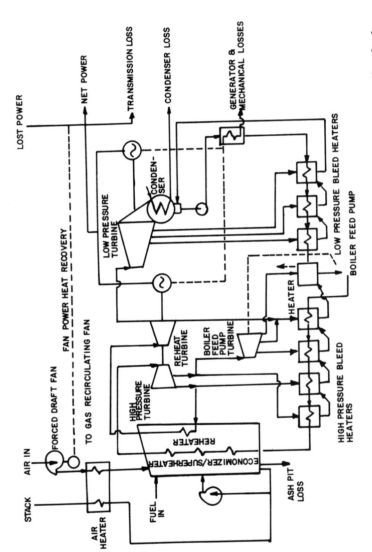

FIG. 5.8 The power cycle of a fossil fuel–single preheat, eight–stage regenerative feed system. The system shown is rated at 3515 psia and 1000° F steam. (From Ref. 127.)

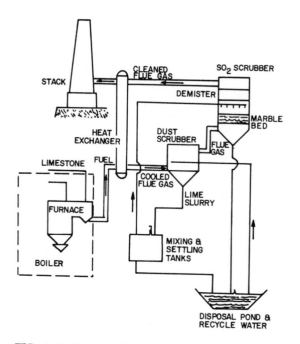

FIG. 5.9 The wet limestone process for removal of elemental sulfur from stack gases. (From Ref. 127.)

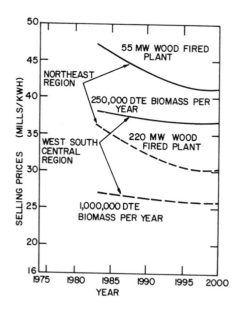

FIG. 5.10 Comparison of market- and production-oriented bus bar selling prices for electricity for wood-fired and coal-fired systems. Bliss and Blake (124) made the comparison based on sites in New England and Louisiana.

and plastics formulations), although it may be a viable approach to large-scale central pulverized wood-fired steam generation.

Bliss and Blake (124) have compared bus bar prices for electricity produced entirely from the combustion of wood biomass in 55 and 220 MW plants to a 1000-MW coal-fired plant, based on two different regions of the country over the period 1975 to 2000. Figure 5.10 shows this comparison. It can be seen from these economic projections that small wood-fired power plants in comparison to large coal facilities do not appear competitive even by the year 2000. At present, standing forests in the Pacific Northwest, if used as a biomass source for fuel in 220-MW power plants are almost competitive with large coal-fired installations. The comparisons shown in Fig. 5.10 does not preclude the use or potential of wood biomass in future electrical generation but does indicate that considerable developmental effort is necessary in order for it to play an important role in this area.

WATERWALL INCINERATION

Incinerators were originally devised for the disposal of refuse and various undesirable combustible materials. The use of raw refuse as a fuel for steam generation in waterwall incinerators has its roots in Europe. Today, such an approach is considered to be accepted technology in the United States. Table 5.2 lists the capacities, location, products, and processing steps used in existing and planned waterwall incinerators in the United States.

Usually refuse is combusted as received and size reduction is only used on oversized or bulky materials. Firing can be accomplished in a number of ways; however, the traveling grate scheme is most often used. During incineration, combustion gases exchange heat in the boiler section, superheater, and economizer, which results in a reduction in flue gas temperature. Flue gas exit temperatures and excess air levels are normally higher than for suspension firing. In addition, boiler efficiencies are lower than in electric utility boilers. Auxiliary fuels such as oil or coal can be used for supplemental steam generation.

INCINERATION OF SOLID WASTES AND
ENVIRONMENTAL CONTROL

The use of incineration for the disposal of solid wastes has been practiced in Europe for many years. Cheremisinoff and Morresi (15) discuss European steam-producing incinerators in depth.

In Saugus, Massachusetts, a 1200-ton/day solid-waste-burning steam generating plant is operating which is expected to produce about 2×10^9 lb of steam a year. The system will service some 16 communities of the north Boston area for refuse disposal and power generation.

TABLE 5.2 Waterwall Incinerator Systems in the United States

Location	Capacity (mil gal/day)	Products	Processing Steps
Connecticut (New Haven)	1632.9	Steam	—
Florida (Dade County)	2721.5	Steam, aluminum, glass, ferrous	Wet pulping, magnetic separation, mechanical separation
Illinois (Chicago)	1451.5	Steam	None
Kentucky (Lexington-Fayette Urban County Gov.)	952.5	Steam, ferrous	Shredding, magnetic separation
Massachusetts			
Braintree	217.7	Steam, ferrous	Magnetic separation
Haverhill	2721.5	Steam, ferrous	—
Saugus	1088.6	Steam, ferrous	Screening, magnetic separation
Michigan (Detroit)	2721.5	Steam	—
Minnesota (Minneapolis-St. Paul)	1088.6	Steam	—
New York (Onondaga County)	907.2	Steam, ferrous	—
Ohio (Akron)	907.2	Steam, ferrous	Magnetic separation, shredding, air classification
Pennsylvania (Harrisburg)	653.2	Steam, ferrous	Magnetic separation
Tennessee			
Memphis	1814.4	Steam	—
Nashville	653.2	Steam	None
Virginia (Norfolk)	326.6	Steam	None

Source: From Ref. 127.

TABLE 5.3 Representative Elements in Municipal and
Commercial Refuse

Major Constituents	Minor Constituents	
Aluminum	Antimony	Manganese
Calcium	Arsenic	Mercury
Chlorine	Barium	Molybdenum
Iron	Beryllium	Nickel
Magnesium	Bismuth	Niobium
Phosphorus	Boron	Platinum
Potassium	Cadmium	Rubidium
Silicon	Cesium	Selenium
Sodium	Chromium	Silver
Sulfur	Cobalt	Strontium
Titanium	Copper	Tantalum
Zinc	Germanium	Tin
	Gold	Tungsten
	Lead	Vanadium
	Lithium	Zirconium

Plants such as this must be designed to meet strict and enforced environmental regulations. Trace elements and various organic constituents present in refuse can result in emissions into the environment during the combustion process. Table 5.3 lists various elements known to exist in refuse. A large number of these substances are potentially hazardous and, as such, refuse is not considered a clean fuel in comparison with agricultural and silvicultural biomass.

Uncontrolled particulate emissions from waste conversion systems, for example, waterwall incinerators, are typically in the range of 2517.2–2974.8 mg/nm^3 (127). Particulate emissions from such systems are comparable to coal combustion and waterwall incineration on a grain loading basis. Trace elements such as arsenic, beryllium, mercury, silver, cadmium, tin, and zinc have been found in fly ash from coal and combination coal/waste systems.

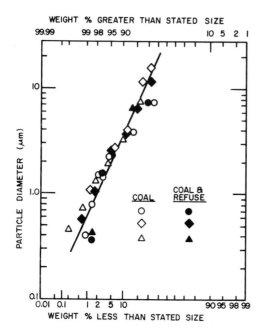

FIG. 5.11 Data on mean particle diameters for uncontrolled emissions from coal-fired and combination coal/refuse-fired systems. (From Ref. 129.)

Stack emissions from wood biomass combustion generate both gaseous and particulate pollutants. Gaseous pollutants include carbon monoxide, sulfur oxides (SO_X), nitrogen oxides (NO_X), and various unburned hydrocarbons. Wood and bark generally contain negligible amounts of sulfur, and so SO_X is not a major problem as it is in refuse incineration.

Particulate emissions from boilers using wood vary and depend on the degree of char reinjection, boiler type, amount of excess air used, wood species, and physical properties of the biomass as well as moisture content.

Data on particle sizes for particulate emissions from municipal incinerators and combined coal/refuse fired systems have been obtained by various investigators. Figure 5.11 shows data obtained from a study by Gorman et al. (129) on coal-fired and coal/refuse-fired incinerators. Fly ash particulates from hog-fuel boilers are relatively light in density, and average particle sizes tend to be larger than fly ash from coal-fired boilers.

Combustion of wood biomass will generate particulate pollutants in the form of emissions and ash. Ash is in the form of captured fly ash and bottom ash. As already noted, unlike coal or oil, wood has almost negligible sulfur content and thus SO_2 is generally not a problem. The same observation can be made for nitrogen oxide emissions from wood burning.

TABLE 5.4 Typical Emission Factors for Wood and Bark
Combustion in Boilers

Pollutant	Emission (g/kg)	Comments
Particulates	12.5-15.0	Atmospheric emission factor without fly ash reinjection
Sulfur oxides (SO_x)	15.0-17.5	For boilers with reinjection (100% reinjection)
	0-1.5	The higher values generally apply to bark
Carbon monoxide	1.0	—
Hydrocarbons	1.0	Principally methane
Nitrogen oxides (NO_x)	5.0	

Source: Data from Refs. 126 and 127.

Emission factors for wood and bark combustion are given in Table 5.4.
Values are for uncontrolled emissions and tend to suggest that they should
not pose major air pollution problems.

Char reinjection systems return collected particulates to the boiler's
combustion zone in order to achieve more complete combustion of carbon.
Reinjection increases boiler efficiency and minimizes the emission of
uncombustibles. Unfortunately, it also increases boiler maintenance requirements, decreases average fly ash particle sizes, and increases dust
loadings in the collector which makes particulate removal more difficult.

ENERGY CONVERSION OF SOLID WASTES

Estimates show that roughly 4.5 billion tons of municipal, industrial,
mineral, and agricultural solid wastes were produced in the United States
in 1970 alone (130). Of this amount, about 13% (approximately 570 million
tons) represented dry combustibles. Table 5.5 gives a breakdown of this
13%.

The fuel value associated with these waste estimates represents an
annual waste of 8.5×10^{15} Btu, or roughly 12% of the total energy requirement of the United States. Not all of this combustible solid waste is readily
available for use as a fuel, however. Although urban-generated waste
may qualify as a fuel because it is available in concentrated form at appropriate locations, wastes from processing, manufacturing, or agricultural
operations are often generated in widely dispersed areas.

Direct Combustion of Biomass

TABLE 5.5 Quantity and Fuel Value of Dry Combustible Solid
Waste Discarded in 1970

Waste Source	Combustibles Discarded (lb × 10⁹)	Fuel Value (Btu × 10¹⁵)
Refuse and sewage/urban generated waste		
Household and municipal	168.4	1.348
Sewage solids	13.8	0.083
Commercial and institutional	62.1	0.496
Manufacturing plant waste	23.5	0.188
Demolition	7.7	0.062
Total	275.5	2.177
Manufacturing and processing waste		
Wood-related wastes	51.3	0.411
Textiles and fabric wastes	0.6	0.005
Nonfabric synthetic materials	0.9	0.011
Food processing	1.5	0.008
Miscellaneous manufacturing	0.2	0.001
Total	54.5	0.436
Agricultural wastes		
Animal wastes	413.4	2.779
Crop wastes	340.0	2.720
Forest and logging residues	51.7	0.414
Total	805.1	5.913
Total combustible wastes and fuel value	1,135.1	8.526

Source: Data from Refs. 131 and 132.

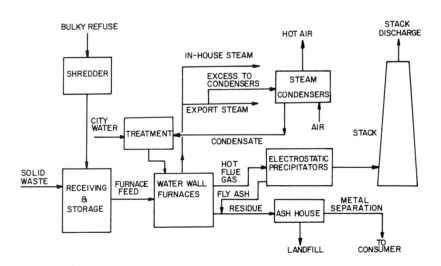

FIG. 5.12 Schematic of the Chicago Northwest Incinerator.

Waste-as-fuel processes are at various stages of development, ranging
from bench and pilot scale through research at the demonstration plant phase
all the way up to commercial applications. Many of these systems parallel
biomass conversion processes.

The recovery of energy from solid waste via waterwall incineration has
been in use on a commercial scale in Europe for several years. The move-
ment of this technology across the Atlantic is gradually being accomplished,
and as systems are being installed in this country, some improvements and
modifications to the basic technology are being made. The following paragraphs present some illustrative projects that utilize this type of technology.

The Chicago Northwest Incinerator. (See Fig. 5.12.) The design is
based on the Martin Incinerator System widely used in Europe and includes
four waterwall furnaces having a combined capacity of 1600 tons/day. Virgin
refuse is taken from a storage area and charged directly to the incinerator
feed hopper. From there, the refuse drops onto a feed chute and then is
fed automatically onto the stoker by means of a hydraulic feed ram. The
solid waste is buried at temperatures of about 1600° F. Secondary air is
added to the flue gas prior to entering the boiler to produce temperatures
in the 2000° F range. The boiler is constructed of waterwalled tubes with
extruded fins. After passing through the boiler, the gases travel through
an economizer section and then into an electrostatic precipitator for particulate removal. Each of the four 400 ton/day furnaces produces 110,000 lb/hr
of 250 psig steam, most of which is condensed in air-cooled condensers.

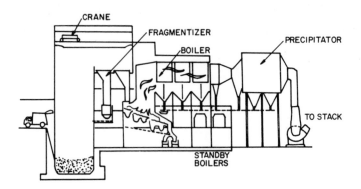

FIG. 5.13 A cross-section of the Saugus plant.

The Saugus (Massachusetts) Refuse Energy Plant. This is another water-wall incinerator. It will initially burn about 1200 tons of solid waste per day from the Boston area while generating some 300,000 lb/hr of 625 psig steam. Figure 5.13 represents a very simplified cross-section of the Saugus operation.

Other full-scale waterwall incinerators that have been installed and operated on this side of the Atlantic include the very large units in Montreal and Hamilton, Ontario, and the Nashville Thermal Transfer unit.

The Horner and Shifrin Fuel Recovery Process (St. Louis). A schematic of this process is shown in Fig. 5.14. Municipal solid waste is prepared for firing in a coal-burning utility boiler. The solid waste provides 10-20%

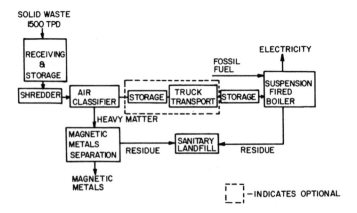

FIG. 5.14 The Horner and Shifrin fuel recovery process.

of the fuel requirement, and coal supplies the remainder. Solid waste is shredded, and the noncombustibles are removed by air classification. The solid waste is then conveyed from a storage bin to a rotary air lock feeder and then pneumatically injected into the boiler.

6

Charcoal Production

CHARCOAL INDUSTRY

Charcoal is the solid residue generated when wood biomass is heated in the absence of air. The charcoal industry began in this country because of the demand for charcoal for pig iron and gun powder. Prior to 1890, U.S. charcoal production closely followed pig iron production because of its use as a reducing agent in blast furnaces.

The U.S. charcoal industry was started around 1620. Iron blast furnaces were constructed in many localities near deposits of iron ore. Iron making did not expand rapidly until after the Revolutionary War and westward migration began.

Early blast furnaces were some 20-30 ft tall with diameters ranging from 4 to 8 ft. They were made of stone and were usually located near steep hillsides so that raw materials could be readily charged through the top. These systems were capable of producing 1-6 tons of iron/day, requiring roughly 1 ton of charcoal per ton of pig iron. By 1850 nearly 377 furnaces were in operation, producing about 563,000 tons/year.

The use of charcoal in iron production began declining when changing technology brought about larger blast furnaces. Charcoal does not have sufficient strength to support the weight of the necessarily large overburden in higher furnaces.

The iron industry generally preferred charcoal that had properties of high crushing strength and could be manufactured from various heavy hardwood species such as maple, beech, oak, and hickory. In the gun powder industry, softer charcoal derived from willow and basswood was preferred.

The hardwood distillation industry began about 1812. Wood biomass was charged in beehive kilns where gases were collected and condensed. Products generated were charcoal, crude pyroligneous acid, and various noncondensible gases. Pyroligneous acid was refined to acetate of lime,

methanol, and tar. Tar and charcoal were both used as fuel. The noncondensible gases were fed back to the kiln and used to preheat the wood.
Beehive kilns were replaced during the latter part of the nineteenth century
with externally heated steel retorts.

The latter part of the nineteenth century brought further expansion to
the charcoal industry with the production of synthetic dyes. The synthetic
dye industry began in 1856 in England. From about 1870 to 1910, over
1300 new dyes were made. Large-scale production of methanol was made
possible by externally heated steel retorts. The use of methanol as a solvent
in the dye industry along with the increasing demands for acetic acid, ethyl
acetate, formaldehyde, and acetone brought about tremendous growth of the
wood distillation industry.

In the mid-1920s, a synthetic process for the production of methanol
was commercialized in Germany. This, and the recovery of acetone from
butanol fermentation, marked an end to the rapid growth of the charcoal
industry. Charcoal production and demand declined until the 1950s, when
the demand for a smokeless recreational cooking fuel opened a new market.
The advent of charcoal briquettes led to increased wood residue utilization.
Charcoal production today is based on wood scavenger prices. Almost any
type and form of wood and bark can be used in its production.

Briquettes are manufactured by crushing the charcoal in a hammermill
and preparing a slurry in water and a starch binder. Approximately 8%
of starch is required for the process. The briquettes are formed in a
double-roll press.

Green briquettes have about 35% moisture content. They are first dried
and then bagged. A typical briquette plant requires roughly 1-3 tons/hour
of charcoal to operate economically.

Older charcoal production with chemical recovery consisted of hardwood
distillation as already described. There was also a softwood distillation
industry in the South which was based on southern pine lightwood (sometimes
called fatwood). Products were charcoal, pine tar oil, refined wood turpentine, and pine tar.

Today, liquid wood smoke for food flavoring is manufactured by the
destructive distillation of hickory and other selected hardwoods. The
process consists of water-scrubbing the wood distillate. Water is recycled
until a certain acid level is reached, and then it is allowed to settle to
precipitate benzpyrene compounds. Foods labeled as smoke flavored are
usually flavored with liquid wood smoke.

There is also an active carbon product that can be manufactured from
wood; however, little is made directly from wood charcoal today. Most
is derived from petroleum sources. Wood charcoal can be activated with
steam, with yields based on dry wood in the range of 12-15%.

There are seven activated carbon plants in this country that have a
capacity of around 290 million pounds per year. A typical wood charcoal
analysis is 20-25% volatiles and 75-80% fixed carbon on a moisture- and

ash-free basis. A typical heating value for charcoal of a good grade is around 12,000 Btu/lb.

Specialized charcoal is derived from other biomass such as nut shells. Such charcoals have found limited use as specific gaseous and liquid constituents in various chemical processes and in water purification. At present, large-scale demand for charcoal as an industrial or utility fuel does not exist.

METHOD OF PRODUCTION

The Nichols-Herreshoff furnace process (Nichols Engineering and Research Corporation of Belle Mead, N.J.) is used as the basis for much of the barbecue-oriented charcoal manufacturing in the United States. The process generates both a solid and gaseous fuel. The system is based on the Herreshoff multiple-hearth furnace, used for many years in various roasting operations on minerals and assorted solids.

Figure 6.1 illustrates the Nichols-Herreshoff multiple-hearth furnace. The system shown consists of a series of disk- and doughnut-type hearths arranged alternately in a vertical fashion. Biomass moves to the center of the top hearth, which is a doughnut type, where it can drop to the hearth at the next level (a disk type). From there solids flow outward to the periphery, where they drop to the hearth at the next level (doughnut). Rotating rabble arms attached to a central shaft extend over each hearth and promote the radial flow of solids inward or outward. The shaft is hollow and cooled via air blown through and then into the furnace for combustion of the fuel. Combustion air can be bypassed at each hearth level for reasons of temperature control.

Figure 6.2 illustrates the process by which charcoal is principally manufactured today. Combustible gases are generated in excess of the amounts required to dry the wood biomass feed. Combustibles can either be burned in a flare, used locally as a fuel gas, or combusted for waste heat steam generation. A typical heating value for the combustible gases generated is on the order of 200 Btu/SCF (standard cubic feet) (124). This is augmented by the contained heat; corresponding to temperatures in the range of 800-1200° F.

A production rate of 2 tons/hr of charcoal generates roughly 50,000 lb/hr of high-pressure steam by-product. Roughly 8 tons of wet wood biomass at 50% moisture content is necessary per ton of charcoal manufactured. The volatile content of charcoal may be adjusted in the process over a range of 10-20%. This content is relatively stable over a range of particle sizes from 8 to 20 mesh.

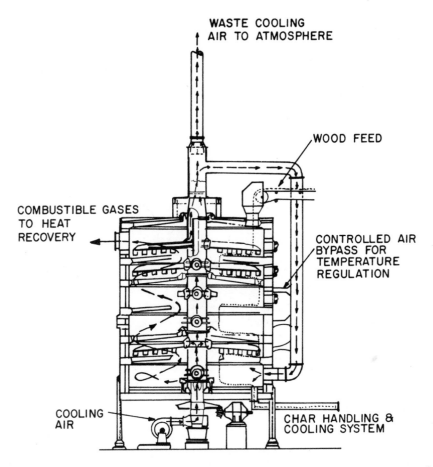

FIG. 6.1 The Nichols-Herreshoff multiple-hearth furnace. (Courtesy of
Nichols Engineering and Research Corp., Belle Mead, N.J.)

CHARCOAL PRODUCTION ON A STAND-ALONE BASIS

Bliss and Blake (124) have estimated production-oriented selling prices for
charcoal manufactured as a separate production effort. They used the
Nichols-Herreshoff process in its conventional mode of operation as the
basis of their analysis, although charcoal can be produced from wood biomass
in conjunction with the production of either fuel oil or fuel gas. Combustion
gases are assumed to pass first through a waste heat recovery unit to generate steam for commercial or utility plant use and then through a feedstock

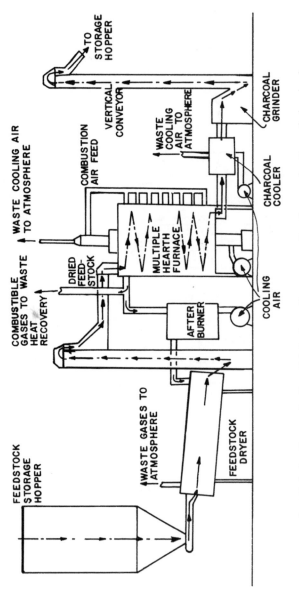

FIG. 6.2 The Nichols–Herreshoff multiple–hearth process for the production of charcoal.

TABLE 6.1 Estimated Production-Oriented Selling Prices of Charcoal Alone and a Mix of Charcoal and Steam

Plant Size (ODT/D)	Wood-Biomass Feedstock Cost ($/10^6 Btu heating value)			
	1.00	1.50	2.00	2.50
Charcoal/steam ($10^6 Btu)				
850	2.08	2.78	3.48	4.27
1700	1.93	2.62	3.33	4.02
3400	1.88	2.52	3.22	3.90
Charcoal alone ($10^6 Btu)				
850	4.90	6.53	8.23	9.88
1700	4.55	6.20	7.90	9.50
3400	4.30	6.00	7.65	9.25
Charcoal alone ($/ton)				
850	108	144	181	218
1700	100	137	174	209
3400	95	132	169	204

Source: From Ref. 124.

drier. Hot char exiting the furnace is cooled, then ground to specification, and conveyed to a storage area.

Three different plant sizes were studied. Total capital investment requirements and annual operating and maintenance cost estimates are given in Table 6.1. Note that cost projections were based on a charcoal heating value of 11,000 Btu/lb.

NEW TECHNOLOGIES

Charcoal contains the bulk of the carbon that is originally present in the biomass. The basic reaction used in the production of charcoal from wood biomass is carbonization by pyrolysis. Depending on the conditions of the carbonization reaction, charcoal may contain volatile materials that can be driven off upon further heating in the absence of air. Volatile gases gener-

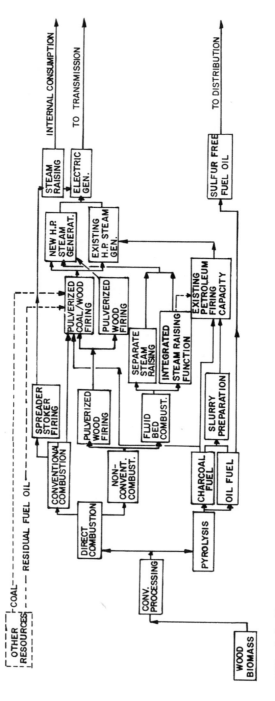

FIG. 6.3 Alternate conversion routes for producing energy from biomass.

ated are combustible and are used for the purposes of igniting and supporting the combustion of charcoal.

Charcoal is comparable to coal both in appearance and chemical composition. Volatile materials present in charcoal generally constitute less than
20%, whereas for coal they are on the order of 20-30%. Charcoal has an
extremely low sulfur content in comparison to nearly all coals.

Two relatively new technologies are being considered aimed at large-
scale use of charcoal in energy production. One approach involves slurrying
pulverized wood charcoal with high-sulfur fuel oil and combusting the slurry
in existing steam-generating equipment primarily designed for fuel oil combustion. The other approach involves mixing wood charcoal with high-sulfur
coal, pulverizing the mixture, and combusting in existing equipment designed
for pulverized coal firing.

Figure 6.3 illustrates how these approaches might fit in with various
energy conversion alternatives discussed thus far.

In the charcoal/oil slurry combustion process, at present only at the
planning stage, sulfur-free charcoal and oil products from pyrolysis can
be slurried with high-sulfur petroleum fuels. The mixture can be used as
a fuel in conventional oil-fired central station boilers. This approach would
(a) reduce reliance on imported petroleum; (b) aid in compliance with existing and possible future sulfur oxides emission regulations; and (c) minimize
the need for aftertreatment of stack gases.

Charcoal mixed with coal prior to pulverization and combustion might
be a viable approach as well. The economics may become highly attractive
once a market has been established for the oil or gas coproducts of manufacture. Mix proportions may be adjusted in order to limit the SO_x emission
standards and to minimize stack gas treatment.

7

Methane Generation
from Biomass

OVERVIEW OF BIOMASS TO METHANE

If organic matter is decomposed under oxygen-free conditions, a flammable
gas, methane, will be generated. Also known as marsh gas, this phenomenon has been observed for centuries. Methane was first recognized as
having practical and commercial value in the 1890s in England, where a
specially designed septic was used to generate the gas for the purpose of
lighting streets. This approach has been used numerous times when conditions of reduced energy supplies existed. For example, during World
War II methane was produced to fuel automobiles in Europe.

Methane generation has also been successfully applied to meeting energy
needs in rural areas. In India, for example, methane-generating units and
plants using cow manure have been in operation for years. In Taiwan, more
than 7500 methane-generating devices utilizing pig manure have been constructed. Some other countries where such devices are in operation are
Korea, where more than 24,000 units were installed between the period
1969 and 1973, and China, Tanzania, Uganda, and Bangladesh (133).

In the United States there has been considerable interest in the process
of anaerobic digestion as an approach to generating a safe, clean fuel, as
well as a source of fertilizer.

The raw materials used in commercial methane generation have been
traditionally classified as waste materials, which include crop residues,
animal wastes, and various urban wastes.

Figure 7.1 illustrates the basic concept behind methane generation.
This approach is advantageous in that it provides:

1. Generation of a storable energy source
2. Production of a stabilized residue that can be used as a fertilizer
3. An energy-efficient means of manufacturing a nitrogen-containing
   fertilizer

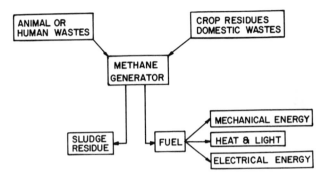

FIG. 7.1 Application of anaerobic fermentation in the generation of methane from organic wastes.

4. A process having the potential for waste sterilization, which can reduce public health hazards from fecal pathogens
5. If applied to agricultural residues, a reduction in the transfer of fungal and other plant pathogens from one year's crop to the next

The biogas system consists of anaerobic digestion of biologically degradable organic material. The amount and quality of gas produced depends on the biomass used. Roughly an upper limit of 8-9 ft^3 of gas can be generated per pound of volatile solids charged to the digester. The product gas will contain anywhere from 50 to 70% methane. Generally, the organic matter must be highly degradable in order to achieve such yields. Lower gas production rates will result from less biodegradable wastes.

Gas generated from the digestion of organic waste is colorless, flammable, and contains roughly 30-40% CO_2 and trace amounts of hydrogen, nitrogen, and hydrogen sulfide. Methane is a nontoxic gas possessing a slight odor.

THE MICROBIAL PROCESS

The anaerobic digestion or organic wastes is often visualized as a three-stage process (refer to Fig. 7.2). The first stage consists of facultative microorganisms attacking the organic matter. Polymers are transformed into soluble monomers through enzymatic hydrolysis. These monomers become the substrates for the microorganisms in the second stage where soluble organics are converted into organic acids. Soluble organic acids, consisting primarily of acetic acid, form the substrate for the third stage. In this last step, methanogenic bacteria, which are strictly anaerobic in nature, can generate methane by two different routes: one is by fermenting acetic acid to methane (CH_4) and CO_2, whereas the other consists of reduc-

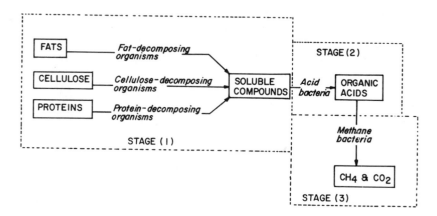

FIG. 7.2 The three stages of anaerobic fermentation of organic wastes.

ing CO_2 to CH_4 via hydrogen gas or formate generated by other bacterial species. Biological growth takes place during all three stages of the fermentation process.

Substrates generated in the first stage will consist of a variety of substances composed mainly of carbohydrates, lipid, protein, and organics. Typical components found in sewage sludge are listed in Table 7.1 (134). The carbohydrates primarily consist of cellulose, hemicellulose, and lignin. Bacteria having the capability of solubilizing these materials must display cellulolytic, lipolytic, and proteolytic enzymatic capacity.

Cellololytic activity reduces the complex raw substance to simple, soluble, organic components. As shown in Table 7.1, the largest portion of organic matter in sewage sludge consists of cellulose. For crop residues used as feedstock in the process, the cellulose content is higher.

TABLE 7.1 Various Components Found in Sewage Sludge

Component	Percentage on a Dry Weight Basis
Cellulose	34.5
Hemicellulose	6.0
Protein	19.0
Lipids	14.0
Ash content	34.0

Source: From Ref. 134.

Cellulose has the basic structure of polymerized glucose units formed in a chain of indefinite length and complex branching networks. Bacteria having cellulolytic capacity reduce the chains and branches first to dimeric and then to monomeric sugar molecules. These are then converted to organic acids.

Cellulolytic bacteria are classified into two groups: mesophilic and thermophilic. The former have an optimal temperature range for digestion between 30° and 40° C. Each group has an optimum pH range of about 6.0-7.0; however, during the initial fermentation process, the organic acids produced cause a lowering of the pH, and it is often necessary to buffer the slurry with lime. An equilibrium pH of 7.0 is reached when the chemical activities of the acid-forming bacteria are in balance with those of the methanogenic bacteria.

During the first stage, interaction between several species of cellulolytic and hydrolytic bacteria promote the decomposition of organic material. Several investigations indicate that synergistic activity is involved as cellulolytic bacterial by-products are digested by noncellulolytic bacteria.

Cellulose conversion to monomers is believed to be rate limiting, as the first stage appears to be slower than the later ones. The rate of hydrolysis has been reported to be a function of the substrate, bacterial concentration, pH, and temperature (133-137).

As already noted, the monomeric species generated by the hydrolytic breakdown become the substrate for the acid-generating bacteria in stage 2. Major acids generated during the second phase include acetic, proprionic, and lactic acids. The methanogenic bacteria strains are generally restrictive in their substrate consumption and are believed to utilize primarily, if not solely, acetic acid. There are methanogenic bacteria which can generate bacteria from H_2 and CO_2. These substrates are also manufactured during the carbohydrate catabolism. Methanol is also a potential by-product of carbohydrate decomposition, from which methane can be derived. Other substrates can be utilized as well, for example, formic acid; however, such reactions are usually of minor importance and not always present in anaerobic fermentation.

During the aforementioned balanced digestion process, the optimum pH range is automatically maintained. Volatile organic acids generated during stage 1 will tend to depress the pH; however, the destruction of volatile acids and formation of bicarbonate buffer during stage 2 will tend to offset this action. An imbalanced condition can, however, arise when the amount of acid produced outweighs the amount of methane generated, causing an excess of volatile organic acids to build up in the system. This can result in a cessation of gas production and a sharp drop in the pH. Several substances which are toxic to methane bacteria can cause this drop or termination of gas generation. Major deterrents include ammonia, ammonia ion, soluble sulfides, and salts of metals. Table 7.2 lists several common toxins that are harmful to methane bacteria and their limiting concentrations (138).

TABLE 7.2 Common Toxins Harmful to Methane Bacteria

Toxin	Harmful Concentration (mg/liter)	Harmful pH
Ammonia (as ammonia nitrogen)	1500-3000	7.4
Ammonium ion	3000	—
Sulfides (soluble)	50-200	—
Metal salts (Cu, Zn, Ni)	—	—

Source: From Ref. 138.

Alkali and alkali-earth metal salts (for example, sodium, potassium, calcium, or magnesium) can be either helpful to or inhibit gas production, depending on the concentrations and forms. The toxicity is generally associated with the cation portion of the salt. Only the constituents in solution represent potential toxins. Therefore, when such constituents are removed from solution, they cannot disturb the organism's metabolism.

Bacteria can become acclimated to toxins over varying concentrations given the proper conditions. Tolerance levels to soluble sulfides and ammonia can be extended by increasing concentrations of these constituents gradually and allowing acclimatization. In general, soluble sulfide concentrations in a digester depend on the incoming sources of sulfur, pH, rate of gas generation, and the presence and availability of heavy metals that participate in the reactions.

Ammonia is usually formed in the third stage from degradation of urea or protein. Ammonia, a source of nitrogen for anaerobic bacteria, is stimulatory to the biological reaction. However, at high concentrations its effects may be toxic to microorganisms.

Ammonium ions exist in equilibrium with dissolved ammonia gas:

$$NH_4^+ \rightleftarrows NH_3 + H^+ \tag{7.1}$$

Soluble ammonia gas is inhibitory at considerably lower concentrations than the ammonium ion. At low values of pH, the equilibrium is shifted to the left-hand side of Eq. (7.1) so that inhibition is proportional to the ammonium ion concentration. Toxicity, as noted in Table 7.2, occurs generally above 3000 mg/liter. At higher values of pH, the equilibrium shifts toward the right-hand side of Eq. (7.1), where if dissolved ammonia is between 1500 and 3000 mg/liter, gas retardation will occur.

Various organics will also cause digestion to be inhibited, e.g., various alcohols at high concentrations. At low concentrations they are often degraded naturally or can be treated out of solution via precipitation. Figure 7.2 reviews the three stages of anaerobic fermentation of organic wastes.

TABLE 7.3 Various Nonmethanogenic Bacteria Isolated from Anaerobic Digesters

| | Isolated on: | | | | |
| | | | Protein | | |
Bacterium	Cellulose	Starch	Peptone	Casein	Lipid
Aerobacter aerogenes					
Alcaligenes bookerii					X
A. faecalis	X				
Bacillus sp.					
B. cereus var. mycoides		X		X	
B. cereus	X	X	X	X	
B. circulans			X		
B. firmus			X		
B. knelfelhampi					
B. megaterium	X	X	X	X	X
B. pumilis			X	X	
B. sphaericus			X	X	X
B. subtilis			X	X	X
Clostridium carnofoetidum	X				
Escherichia coli			X	X	
E. intermedia					
Micrococcus candidus		X			

(continued)

Table 7.3 (continued)

| Bacterium | Isolated on: | | | | |
| | Cellulose | Starch | Peptone | Casein | Lipid |
			Protein		
M. luteus					X
M. varians		X	X	X	
M. ureae		X			
Paracolobacterium intermedium			X		
P. coliforme			X		
Proteus vulgaris	X				
Pseudomonas aeruginosa	X				
P. ambigua					
P. oleovorans					X
P. perolens					X
P. pseudomallei					
P. reptilivora	X				X
P. riboflavina	X				X
Pseudomonas spp.	X	X	X	X	
Sarcina cooksonii					
Streptomyces bikiniensis					X

Source: Data from Refs. 139 and 140.

The major anaerobic bacteria and various materials which they digest are listed in Table 7.3 (139, 140). Note that operation and design of methane-generating systems will vary with the type of residue being digested. For example, cellulose digesters are likely to be used for converting urban solid wastes to methane, as the digestible organic fraction of such residue is predominantly cellulose.

As noted, methane-forming bacteria are sensitive to variations in their environmental conditions. Under optimum conditions, the growth rate of a microbial population appears to follow a geometric progression with time. That is, individual cells will multiply and divide at roughly the same rate as their parent cells did. Thus, optimally the number of bells doubles with each new generation.

Figure 7.3 illustrates a typical bacterial population growth curve. The exponential growth period follows an initial lag phase. After the exponential expansion phase, nutrients gradually become depleted and toxic metabolites begin to accumulate. At this point, the number of viable bacteria becomes constant; this plateau is called the maximum stationary phase. Ideally, in the anaerobic digester the rate-limiting bacteria (i.e., the methane producers) should be maintained in the maximum stationary phase for as long as possible. This is the stage at which maximum gas production takes place. During this phase, population growth continues; however, its rate is counterbalanced by an equivalent death rate. Once the death rate exceeds the growth rate, the culture begins to die exponentially.

The amount of time necessary to reach the maximum stationary stage after a change in environmental conditions has occurred is a parameter requiring further investigation. With the use of seed cultures, steady state operation has been reached within 2 to 3 months.

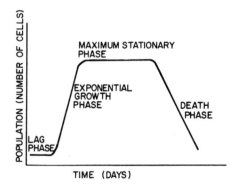

FIG. 7.3 Typical bacterial growth curve.

During the gas production phase, equal amounts of CH_4 and CO_2 may be generated according to the equation:

$$n(C_6 H_{10} O_5) + nH_2 O \rightarrow n(C_6 H_{12} O_6) \rightarrow 3nCH_4 + 3nCO_2 \qquad (7.2)$$

Equation (7.2) describes the anaerobic digestion of carbohydrates such as cellulose. Small amounts of NH_3, $H_2 S$, H_2, N_2, and O_2 will also be found in the gaseous product.

The gas composition is a strong function of pH and temperature. The reactor detention time also plays a significant role but to a lesser extent. Shorter detention times generally produce greater amounts of CO_2, which will be washed out of the system as fresh liquid is charged. Carbon dioxide will reduce the heating value of the product gas as it dilutes the methane (complete CO_2 removal can increase the digester heating value to nearly 1000 Btu/SCF, as required by natural gas utilities).

The digestion process must be designed to operate at constant temperature conditions. Relatively slight to moderate variations in temperature can deter growth patterns of the methane producers. Temperature fluctuations can cause an imbalance between bacteria and acid-producing strains, result-ing in a build-up of volatile acids. McCarty (141) studied gas production as a function of temperature. Figure 7.4 illustrates the effect of tempera-ture on gas production. As shown in the plot, two maxima are observed—one at about 40° C, which corresponds to gas production by mesophilic methane producing organisms; and one at 55° C, corresponding to thermo-philic methane production. Temperatures in excess of 65° C cause gas production to slow down and eventually stop.

When a unit is operating in the range of thermophilic temperatures, an increase in the metabolic rate is observed, resulting in a higher rate of gas production. This implies that shorter retention times are possible.

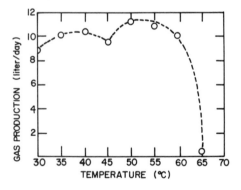

FIG. 7.4 Effect of temperature on gas production, as studied by McCarty (141). Two maxima were found.

The primary advantages of operating in this range are (a) a decrease in the net amount of sludge formed; (b) destruction of the more pathogenic organisms; (c) the potential for increased digestion efficiency; and (d) rapid generation of the organism population occurs after scouring.

Major disadvantages include higher heating requirements and higher volumes of CO_2 produced in the digester gas than in the mesophilic temperature range.

Advantages of operating in the mesophilic range include the following: (a) less water vapor is produced in the gas than when operating in the thermophilic range; (b) less CO_2 is generated; more types of bacteria grow and produce methane; and (d) heating requirements are lower.

Pfeffer and Liebman (142) noted that even small variations in temperature can cause a marked change in the digestion process. An accidental temperature increase of roughly 2.5°C in the thermophilic range (from 60° to 62.5°C) has been observed to lower methane production; after correction, 1 week was required to regain the former rate of methane production in Pfeffer and Liebman's (142) experiments.

Temperature has a fairly complex effect on the system. In particular an optimum temperature must be maintained in order to establish the bicarbonate equilibrium for a given pH. The ratio of bicarbonate to CO_2 in solution is critical, as the amount of carbonic acid in solution will increase with increasing temperature. Bisselle et al. (139) have noted that for urban trash methanation at a pH of 7.0 and a temperature of 30°C this ratio is about 4.9, whereas at 60°C the ratio increases to roughly 5.2. Reduction of soluble CO_2 at higher temperatures is believed to alter the

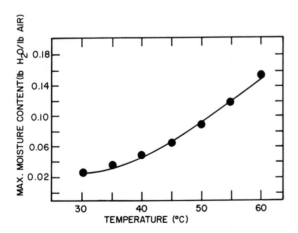

FIG. 7.5   Effect of temperature on the moisture content of saturated air. (From Ref. 143.)

balance between CO_2 in solution and in the gaseous phase, with the net result of a decrease of CO_2 in the digester gas.

Higher temperatures do, however, cause a decrease of CO_2 solubility. For example, in going from 30° to 60°C, CO_2 solubility decreases by a factor of 0.85. Such a reduction causes an increase of CO_2 in the gaseous phase which outweighs the slight shift in the CO_2-bicarbonate reaction toward increased conversion to bicarbonate. At any pH, increasing temperature causes a lower alkalinity as the loss of CO_2 from the liquid to the gaseous state decreases the bicarbonate concentration.

Temperature will also impact on the amount of water in the gas phase. The effect of increasing temperature on water vapor pressure and on the maximum moisture content of the gas at atmospheric conditions is shown in Fig. 7.5 (143). This indicates that more expensive methane-generating systems will be necessary when operating in the thermophilic temperature range as the additional step of drying the gas produced will be necessary.

OPERATING PARAMETERS

In addition to temperature, a number of other parameters must be carefully controlled to ensure proper operation. The methane producers used are strictly anaerobic; consequently, the reactor must be carefully designed to maintain these conditions. Start-up is critical to the operation. Head-space should, for example, be purged with nitrogen or with gas generated in a sister digester. Casida (144) points out that a floating lid is sometimes employed in sewage sludge digesters to maintain anaerobic conditions. In this arrangement the tank lid floats on the liquid, thus accommodating changes in volume. It is possible to pipe gases from the floating lid; however, this is generally considered to be an expensive approach. In addition, the design presents a variety of problems, including difficulties in providing adequate sealing.

The process can be operated over a pH range of 6.2-7.6. An optimum pH of about 6.8 has been observed by several experimenters, although slightly higher pH levels have produced increased gas production. Values below pH 6.2 generate acid conditions that are generally so toxic to bacteria that digestion tends to stop.

Microorganisms are highly sensitive to pH changes. Buffering is necessary for pH control and is therefore an essential step in the overall operation. Natural buffering exists in the system, but additional buffer is often necessary. In sewage sludge digestion, lime is primarily used for this purpose.

Bisselle et al. (139) note that to maintain pH near neutral conditions at 40°C requires a bicarbonate alkalinity, in the form of $CaCO_3$, of a minimum of 1.5 g/liter in a 30% CO_2 environment. A minimum of 1.0 g/liter is necessary at 60°C to maintain the same pH and CO_2 levels. To counteract pH drop caused by excess volatile acids, alkalinities in the range of 3-4 g/liter may be necessary (139).

Another control parameter is the level of nutrients. Sewage sludge generally contains sufficient amounts of nutrients; however, this may not be the case for all types of biomass. For example, urban solid wastes may require nutrient addition as wastes may be deficient in nitrogen and phosphorus. The addition of these nutrients are essential to microbial growth and gas production. The addition of sewage sludge to solid waste has been suggested as a means of supplying additional nutrients.

Sludge actively involved in the anaerobic digestion process in a treatment operation is generally a richer source of nutrients and microbes than is raw sewage. Such material generally has greater buffering capacity than solid waste because of its protein content. The benefits derived from the addition of sludge to urban solid wastes will depend on: (a) waste composition; (b) sludge characteristics; and (c) the proportions of waste and sludge in the digester.

As noted earlier, toxicity levels will also have to be carefully monitored. In the case of solid wastes, it may be extremely difficult to monitor and identify toxic materials with incoming wastes on a continuous basis. Only the soluble materials will be toxic to the microorganisms. Toxicity can be regulated either by removal of materials (i.e., precipitation) or by rendering them inactive through chemical treatment. Still another approach is to dilute the feed biomass stream until levels are below the toxic threshold.

Alkali and alkaline-earth metal salts (e.g., sodium, potassium, calcium, or magnesium salts) are toxic above certain threshold concentrations. Table 7.4 gives typical toxicity levels and tolerable limits for various compounds (145, 146). In municipal wastes, salt concentrations are usually well below the toxicity levels. A possible problem might arise for pH control when such compounds are introduced at high concentrations. In addition, low but soluble concentrations of heavy metals (e.g., from copper, zinc, or nickel salts) are also toxic. The toxicity levels of microorganisms must be carefully considered and applied to the selection of materials used in construction of the digestion vessel. Some heavy metals can be precipitated out of solution in the form of sulfides; however, it must be noted that sulfides are corrosive and toxic themselves when present as insoluble metal precipitates.

Ammonia, which is used as a nitrogen source, is also toxic at high concentrations. Ammonia gas dissolved in solution is toxic to organisms at considerably lower concentrations than ammonium ion. The equilibrium relationship is defined by Eq. (7.1).

There are also a variety of organics that will inhibit anaerobic digestion. Included in these are organic solvents, alcohols, and long-chain fatty acids. Pesticides in agricultural biomass may present a significant problem.

Two other major operating parameters are the feed concentration and retention time.

TABLE 7.4 Toxicity and Tolerable Limits of Bacteria
to Various Compounds

Substance	Tolerable Limits (mg/liter)	Toxic Threshold (mg/liter)
Sodium	3,500-5,500	8,000
Potassium	2,500-4,500	12,000
Calcium	2,500-4,500	8,000
Magnesium	1,000-1,500	3,000
Ammonium	—	1,500-3,000 (pH > 7.4)
Sulfides	—	200

Source: Data from Refs. 145 and 146.

Feed Concentration

To achieve a steady gas production, feed solids concentrations must be
properly set. For a constant gas production rate, increasing concentrations
of solids in the feed generally reduces the digester volume requirements.
This consequently lowers digester heating and dewatering costs, but it does
increase power requirements for mixing. Water requirements limit solids
concentrations to under 40% (by weight) for proper bacterial growth. This,
however, is not the only criterion for limiting suspended solids concentration; for example, increases in energy consumption to achieve greater
mixing must be weighed against marginal increases in gas energy production.
Physical, economic, and positive energy balance constraints will play a
role in limiting the maximum suspended solids concentration.

Digesters using sewage sludge are generally charged at a total solids
concentration of 3-10%. Sewage sludge has been operated in laboratory
digesters at solids levels of 37%, but only under conditions where volatile
acid concentrations were below 2000 ppm. At higher solids loadings, the
potential exists for overloading the digester with organics, which will increase levels of volatile acids, CO_2 content of the gas, and the time necessary to reach a steady-state digestion.

The advantage of higher solids loadings are smaller reactor volumes.
If the biomass is dilute, high retention times* are necessary and so relatively large volumes are required and subsequently high heating costs

---

*Note that the terms detention time and retention time are used interchange-
ably in this book.

incurred. For very low solids concentrations, long retention times become economically limiting. Longer retention times give larger costs because of additional volume and heating requirements.

Retention Time

Retention time is dependent on both the solids concentration and the loading rate to the digester. Bisselle et al. (139) give the following expression to describe this relationship:

$$\tau = 62.4 \frac{X}{R} \tag{7.3}$$

where

τ = retention time (days)
R = solids loading (lb/ft^3 per day)
X = fraction of solids in sludge on a dry weight basis

Pfeffer and Liebman (142) and Pfeffer (147) found that cumulative gas production increases asymptotically as a function of time. Figures 7.6 and 7.7 illustrate this observation for both mesophilic and thermophilic temperature ranges.

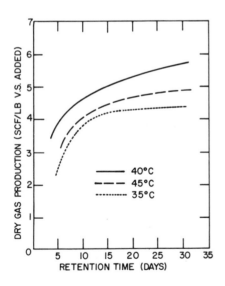

FIG. 7.6 Gas production as a function of reactor retention time for mesophilic temperatures. V.S. = volatile solids. (From Ref. 147.)

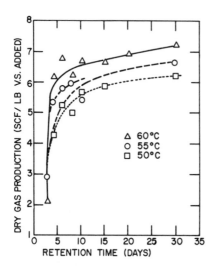

FIG. 7.7 Gas production as a function of reactor retention time for the thermophilic temperature range.

The shorter retention times have been observed to produce the highest gas production on a daily basis in addition to a greater washout of dissolved CO_2. The disadvantage of shorter times is that a smaller portion of the volatile solids are converted to product gas, subsequently generating a larger solid waste disposal problem.

Effluent rates from digesters may also impact on the microbial population. For example, as retention time becomes smaller, a larger portion of active cells may be washed out of the system. At very low retention rates, microorganisms may be washed out at a rate faster than they can reproduce. Minimum retention times will be a function of cell generation rate as well as temperature.

MATERIALS PREPARATION

Table 7.5 lists the potential sources and diverse nature of various biomass forms for methane generation. Such biomass may be residues resulting from other sources or may be directly harvested. The actual feed preparation in a methane-producing system is often accepted as raw biomass which is likely to be quite variable in composition, texture, and moisture content; this material must be converted to a relatively homogeneous feedstock.

Animal manure is the substrate most widely used. It contains considerable amounts of natural fibers that have already undergone both biochemical

TABLE 7.5 Potential Sources of Biomass for Methane Generation

Item	Description
Animal wastes	Cattle-shed wastes (e.g., dung, urine), poultry litter, sheep and goat droppings, slaughterhouse wastes (such as blood, meat), fishery wastes, wool wastes
Crop residue	Sugar cane trash, weeds, corn, and other crop stubble, straw, wasted fodder
Human wastes	Feces and urine (night soil), refuse
Forest residue	Twigs, bark, branches, leaves, undergrowth
By-products/wastes from agricultural related industries	Bagasse, rice bran, tobacco wastes, seeds, wastes from fruit/vegetable processing, press-mud from sugar manufacturing, tea waste, cotton dust (textile industry), oil cakes
Aquatic-based residue	Marine algae, seaweeds, water hyacinths
Solid wastes/urban	Paper wastes, household residues, trash

and mechanical treatment in the animal's digestive tract. This form of biomass requires the least pretreatment prior to methane generation.

Materials of plant origin, on the other hand, vary greatly in composition. In general, water-soluble substances such as sugars, amino acids, proteins, and minerals decrease with the age of the plant. Substances such as cellulose, lignin, hemicelluloses and polyuronides (e.g., pectins, gums, mucilages) increase in content with plant age. The actual nutrient content is rarely low enough to limit the anaerobic digestion rate; however, the biological availability of organic substances is almost always near a rate-limiting level (148). Thus, although sugars may exist as components of cellulose, until the cellulose is hydrolyzed anaerobic digestion and methane production cannot continue. In addition, the greater the concentration of lignin present to protect the cellulose from bacterial action, the less cellulose is available for the digestive process. In an anaerobic fermentation process, lignin is usually considered to be a nonbiodegradable portion of the plant. It is insoluble in water, ether, alcohol, dilute alkali, and about 70% sulfuric acid and as such is not susceptible to bacterial attack.

Paper wastes are another potential source for the anaerobic digestion process. The digestibility of papers is generally quite high, as manufacturing steps have already removed a major portion of the lignin present in the wood. Variations in the feed preparation steps may have significant

effects on digester feedstock and, ultimately, on the overall system efficiency. Pretreatment may affect several characteristics of the digester
feedstock, including: (a) the ratio of digestible to nondigestible materials
in the digester feedstock; (b) the size of the particles in the digester feedstock; (c) the potential presence of digestion-inhibiting substances in the
feedstock; and (d) in the case of urban solid wastes, the fraction of organics
in the feedstock that will eventually be charged to the digester.

To achieve proper particle size variations in order to optimize gas production, shredders may be used. As noted earlier, digestion-inhibiting
substances can greatly damage the digestion process. Highly toxic substances (e.g., pesticides or other commercial poisons) may be present in
municipal waste streams or agricultural-silvicultural biomass. Monitoring
for such substances may require extensive as well as elaborate equipment
and techniques.

Particle size is important not only to digester efficiency but also to
separation efficiency. Furthermore, separation efficiency can be affected
by variations in separation equipment, machine operation, and feedstock
characteristics.

Application of air classifiers has been suggested as one approach to
processing paper-containing wastes, particularly for metal and glass
removal; removal of nonbiodegradable material; and pretreatment for com-
posting, fermentation, or retorting, with a maximum particle size of 2 in.
The removal of nonbiodegradable fine particles from dry waste has been
accomplished by a combination of screening and air classification (139).
Difficulties have been noted in the air classification of the coarser paper-
containing fractions after screening has taken place because of the small
differences in density of the remaining fine particles—paper, cardboard,
and plastic—and the intimate mixture produced by shredding and screening
operations.

The use of conventional hammermills has also been suggested; however,
these tend to tear rather than shred fibrous material. This can result in
a dry pulping, or felting, effect, causing shredded material to agglomerate,
forming flocs. Large flocs will trap and carry a considerable amount of
light and fine material.

Separation of ferrous metals and air classification of nonbiodegradable
fine particles from biodegradable materials are generally more efficient
approaches with larger particle sizes. With respect to urban solid residues,
most food waste, heavy plastic, yard waste, cloth, and wood are heavier
than paper and can be separated from the light organic fraction during air
classification. These forms of biomass make up some 17-32% of all municipal solid waste. Ideally, all plastics should be excluded from the digester,
but the fraction which remains is usually so low that it should have little
effect on digestion. Hence, less separation efficiency may be necessary
with certain types of municipal wastes. Boettcher (149) notes that greater
separation efficiency of the heavier organics can be achieved with larger

particle sizes. For the digester feed to be practical, particle sizes would have to be of sufficient size to counteract felting effects and at the same time small enough to allow for the inclusion of food wastes, yard wastes, and wood in the product from a second stage classification.

Magnetic separation of ferrous metals from urban wastes has been practiced for several years. Large metal wastes such as appliances must be reduced to maximum sizes of about 8 in. in order to be processed by a magnetic separator. Reduction to an extremely small size is not essential; however, it is important that particles be physically detached from each other.

SLUDGE DEWATERING OPERATIONS AND USES

The three most commonly used methods of sludge dewatering are air drying, vacuum filtration, and centrifugation.

Air drying is normally accomplished in sand beds. Moisture is removed by two mechanisms: (a) filtration through sand, and (b) evaporation of the water to equilibrium moisture content. Air drying of sewage sludge reduces the moisture content to roughly 65%. This approach requires substantial land area and does not permit recycling of liquid effluent.

In vacuum filtration, solids are separated from the liquid by means of a porous medium such as cloth, steel mesh, or tightly wound coil springs which retain the solids. Several variables affect filtration rate and final moisture in vacuum-filtered sludges. The major ones are as follows:

1. Pressure drop across sludge and filter
2. The area of filtering surface
3. The initial solids concentration
4. The viscosity of filtrate
5. The size and shape of solid particles
6. The chemical composition of sludge and liquid
7. Chemical conditioners, such as organic polymers, which improve filterability

In centrifugation, the effect of solids concentration must be considered in equipment selection. Not every centrifuge machine is appropriate for dewatering slurries with high solids concentration. In addition, the mechanism of cake discharge must also be considered.

Another factor that will require careful consideration is treatment of the supernatant. Since conditions in the digester facilitate microbial growth, the effluent will have a high level of contamination [e.g., high biochemical oxygen demand (BOD) and chemical oxygen demand (COD), and high concentrations of colloidal or suspended solids].

Suspended solids may be removed by chemical coagulation. For example, colloids can be destabilized by chemicals such as alum or ferric chloride.

Reverse osmosis might be applied to effect removal of dissolved salts.
Ultrafiltration with a high-molecular-weight cutoff membrane would allow
passage of the dissolved salts. Pore sizes would be sufficiently small to
retain suspended and colloidal particles, resulting in a water product with
low turbidity. Supernatant might be recycled into the digestion tank, or be
piped or transported to a sewage plant.

Sludges generated from the anaerobic digestion may have varied use
depending on their biomass source. The organic portion of sludges derived
from plant and animal residue may contain some 30-40% lignin and undigested
cellulose and lipids. The remaining portion will consist of material that
was originally present in the feed biomass. Incomplete decomposition is
attributed to lignin and cutin, newly synthesized bacterial cellular materials,
and volatile fatty acids. Generally, the digested sludge will contain only
small amounts of bacterial cell mass and so odor and insect-breeding problems might not be expected to pose serious problems when sludges are
stored or spread on land. This will of course vary depending on the biomass source. There is a substantial potential for ground and surface water
pollution by the leachate from digester solids used in landfill applications.
Thus, it may be necessary to analyze sludges to determine whether they
contain pathogenic microorganisms.

Application of anaerobically digested sludges to soils has about the same
effect as the application of other types of organic matter. Humus materials
generated usually improve soil properties such as moisture-holding and
water-infiltration capacities. In addition, sludges provide a source of
energy and nutrients which are essential to the development of microbial
populations.

Anaerobic digestion of plant biomass usually conserves the nutrients
that are needed for continued crop production. Organic or ammonium
nitrogen compounds are generally preserved during the anaerobic digestion
of plant residues. [Idnami et al. (150) have shown that roughly 99% of the
nitrogen present in the original biomass is retained after digestion, with
only about 1% lost in product gases.] The distribution of nitrogen between
organic and ammonium nitrogen in the sludge residue strongly depends on
its distribution in the feed material. As only small amounts of nitrogen
contained in plant and animal residue are involved in the anaerobic digestion,
the greater the nitrogen content of the biomass feed, the greater the ammonium nitrogen in the sludge.

Nitrogen losses will generally occur in the handling and storing operations after digestion. For example, improper sludge application to land
can cause nitrogen loss by volatilization. To minimize ammonia-nitrogen
losses, digested matter might be stored in deep lagoons or vessels, which
would offer a minimum surface area for volatilization of the nitrogen in the
sludge.

PLANT DESIGN AND OPERATION

For product gases to have a market value as a fuel, at least 90% must consist of methane and the product must have a heating value of at least 1000 Btu/SCF. In addition, the gas must be low in moisture content. The amount of methane generated from a system will depend on the nature of the biomass. Gas production will also depend on temperature and digestion time.

The reactor (digester) should be designed to provide the best possible growth conditions for methane-producing bacteria. The inner surface should be constructed of materials which will not corrode in the presence of fermentation products or contribute toxic ions to the digestion mixture. The system must be properly designed to provide for the introduction of feed and other additives, withdrawal of effluent, and collection of gas produced. Provision for withdrawal of culture samples during digestion is also desirable. In addition, the capabilities of monitoring pH and operating temperature in the digester as well as adjusting those values during the digestion process should be built in.

Systems should be run on either a batch or continuous basis. Additions should be made daily or so frequently as to be considered semicontinuous.

A batch process would consist of a constant volume process with reactants enclosed in the tank and with no additions or withdrawals of volatile solids or waste products during the retention period. The headspace would be fitted with a valve permitting gases to flow out of the reactor as they are evolved so that a constant headspace pressure may be maintained. For a batch process, the digester would be emptied at the end of the retention period. The advantage of a batch process is that the nature of the feed varies in physical or chemical properties or changes its properties during digestion and mixing. Each batch can be processed for a different amount of time or under different conditions until digestion is satisfactory. A system such as this would require a highly efficient bacterial culturing facility to shorten the time needed to attain a maximum stable gas production level.

In the continuous mode, the feed and additives would be added continuously to the reactor and a portion of the mixture withdrawn at a steady rate. The process might also be carried out on a semicontinuous basis, with feed added after an equivalent volume of reactor contents had been withdrawn.

Christopher (150) experimented with the anaerobic digestion process using simulated solid waste and was able to sustain stable digester operation with additions of small amounts at frequent intervals. The study showed that the optimum time between additions was a function of particle size. Single large daily increments of finely divided solids degraded the digester performance. With larger pieces of solid residue, longer periods between feed additions were possible.

In general, continuous operation appears to be less flexible than the batch mode since all materials passing through the system receive the

same treatment regardless of properties. In addition, it may be difficult
for the front end to provide feed to the digester on a continuous basis.

A semicontinuous process that uses intermittent additions and withdrawals
would have some of the advantages of continuous operation while maintaining
greater flexibility.

Whether the operating plant should be constructed on the site of the biomass source or elsewhere depends on the economic justification for transporting residues. For example, if a system is to be designed to produce
methane from human and animal wastes in a farm or rural area, the biomass
might well be processed at the site. On the other hand, if the system is to
accommodate a village, an off-site installation, away from the residential
area, would probably be more beneficial. One type of plant design for converting night-soil and cattle-shed wastes is illustrated in Fig. 7.8.

Two other biogas plant designs for methane production are illustrated in
Figs. 7.9 and 7.10. Gas generated from each of these units is reported to
be in sufficient quantities to meet the cooking needs of five families (151,
152).

Reactor size will depend on a number of conditions including:

1. The amount of biomass to be processed
2. The type of biomass
3. Average particle sizes
4. Heating requirements
5. Construction materials
6. Degree of mixing

Most of these items have already been discussed. The degree of mixing
may have as large an influence on the digester size and the amount of gas
produced as particle size does. Mixing is necessary to expose new surfaces

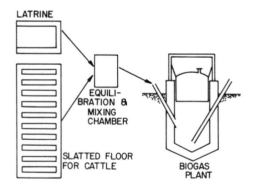

FIG. 7.8 A biogas plant for methane generation. Biomass feed consists
of cattle-shed wastes and night soil.

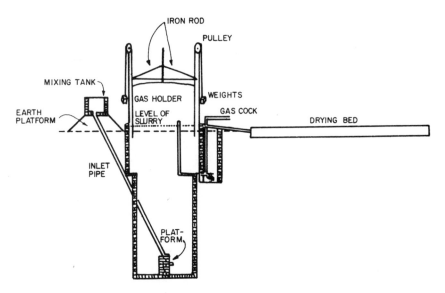

FIG. 7.9 Biogas plant designed for methane production from cow manure. The system was designed by Acharya (151) and developed at the Indian Agricultural Research Institute.

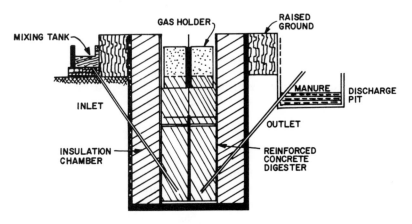

FIG. 7.10 Another biogas plant design, developed by the Gobar Gas Research Station in India. (From Ref. 152.)

to bacterial action. It also deters slowdown of bacterial activity caused by local depletion of nutrients or concentrations of metabolic products. For small reactors, agitation may not be necessary because of relatively long detention times (on the order of 50–60 days) and low gas production rates.
There are a variety of mixing methods. Some of these are listed here:

1. A daily feed to the digester (i.e., semicontinuous operation)
2. Varying inlet and outlet arrangements to promote mixing by influent and effluent flow conditions
3. Use of manually or mechanically operated mixers
4. Creating a flushing action of the slurry through a flush nozzle
5. Flushing the slurry so that the flow is tangential to the digester contents
6. Constructing wooden conical beams that will cut through the scum layer as the liquid surface moves up and down during filling and emptying the reactor

In municipal digester operations, a portion of the product gas is recirculated in order to achieve mixing. There is no reason why such an approach could not be applied to large biogas digester operations.

The actual digester can be constructed either as an aboveground or underground installation. In addition, it may be desirable to operate several digesters in series or to compartmentalize large units. This has the advantage of achieving higher degrees of biomass stabilization as well as increased gas production.

Starting up a digester involves seeding it with an adequate amount of both acid-forming and methane-producing bacteria. Seed material might consist of digested sludge from a municipal digester or perhaps a cow-dung slurry. As a rule of thumb, in a continuous or semicontinuous operation the seed should be at least twice the volume of the fresh manure slurry during start-up. Usually the amount of seed material added daily is decreased over a 3-week period.

The choice of operating temperatures (mesophilic or thermophilic) will depend upon various considerations including climatic conditions. Since methane-producing bacteria are highly sensitive to thermal variations, close monitoring of temperatures will be required.

GAS REFINING

For the product gas to be acceptable as a fuel, in addition to containing a minimum of 90% methane and having a heating value of at least 1000 Btu/SCF, it must have a low moisture content.

Theoretically, carbon dioxide and methane are produced in equal quantities during anaerobic digestion; however, CO_2 has a higher solubility potential than CH_4. Thus, the amount of CO_2 remaining in solution can

TABLE 7.6 Typical Digester Parameters/Variables

Reactor	Retention Time (hr)	pH	Alkalinity (mg/liter CaCO$_3$)	Caustic Added (meq/day)[a]	Percentage Output CH$_4$	Percentage Output CO$_2$	SCF Gas Per Pound Volatile Solid Added
Reactor and gas variables at 60°C							
1	3	6.8	1030	9.14	62	38	1.96
2	4	6.8	1140	7.85	58	42	6.17
3	6	6.8	1530	6.43	57	43	6.75
4	8	6.9	1600	5.57	57	43	6.16
5	10	6.9	1920	3.00	57	43	6.64
6	15	6.8	1900	1.86	56	44	6.61
7	20	6.8	1900	0.86	55	45	6.90
8	30	6.8	2030	0.29	55	45	7.20
Reactor and gas variables at 50°C							
1	3	6.75	1156	92.0	57	43	—
2	4	6.78	1154	87.0	60	40	4.20
3	6	6.83	1251	54.7	61	39	5.25
4	8	6.81	1394	52.5	54	46	4.98
5	10	6.82	1544	45.0	50	50	5.60

(continued)

Table 7.6 (continued)

Reactor	Retention Time (hr)	pH	Alkalinity (mg/liter $CaCO_3$)	Caustic Added (meq/day)[a]	Percentage Output CH_4	Percentage Output CO_2	SCF Gas Per Pound Volatile Solid Added
[Reactor and gas variables at 50°C]							
6	15	6.85	1778	22.5	46	54	5.75
7	20	6.86	2082	15.0	40	60	5.90
8	30	6.89	2343	12.5	35	65	6.20
Reactor and gas variables at 40°C							
1	4	6.77	1244	80.3	63	37	3.6
2	4	6.77	1319	80.3	62	38	3.6
3	6	6.78	1595	70.3	53	47	4.16
4	8	6.78	1780	52.0	48	52	4.3
5	10	6.78	1982	49.0	42	58	4.6
6	15	6.80	2356	36.0	35	65	5.0
7	20	6.83	2554	23.0	35	65	5.3
8	30	6.87	3087	19.0	30	70	5.6

[a]meq = milliequivalents.
Source: From Ref. 147.

be controlled by the digester temperature, retention time, bicarbonate concentration, and pH. These parameters can be optimized within certain limits to reduce the CO_2 concentration in the resultant gas.

Pfeffer (147) studied methane yields and gas composition over a fairly wide range of retention times. Table 7.6 lists a range of digester parameters examined in Pfeffer's study. It should be noted that digester gas moisture content will vary considerably. Pfeffer found moisture content to be a function of temperature. For the thermophilic system at 55°C, moisture levels were as high as 11%; for the mesophilic process at 40°C, a maximum of 5% was reached.

There are a variety of techniques that can be applied in order to remove impurities from the product gas. Scrubbing techniques may be the simplest approach to removing CO_2 and water vapor.

Table 7.7 gives solubilities of CO_2 in water at various temperatures and pressures. Increased pressure tends to reduce water requirements; however, it also introduces corrosion problems in the compressor.

Carbon dioxide can be removed from the digester gas either by wet or dry process scrubbers. Two widely used methods of purifying natural gas employ (1) a bubble column, i.e., the Girbotol process, or (2) molecular sieve absorbents.

The bubble column process is shown in Fig. 7.11. In the system, carbon dioxide and hydrogen sulfide, if present, are absorbed in the aqueous low temperature solution of an ethanolamine compound. This solution then enters a stripping column where the CO_2 and H_2S are released upon heating. The theanolamine compound is then returned to the bubble column to absorb more acid gases.

Three ethanolamine compounds can be used in the bubble column process, namely, monoethanolamine (MEA), diethanolamine (DEA), and triethanolamine (TEA). These compounds have different absorptive capacities for

TABLE 7.7 Solubility of CO_2 in Water (lb CO_2/100 lb H_2O)

Pressure (atm)	Temperature (°F)				
	32	50	68	86	104
1.0	0.40	0.25	0.15	0.10	0.10
10.0	3.15	2.15	1.30	0.90	0.75
50.0	7.70	6.95	6.00	4.80	3.90
100.0	8.00	7.20	6.60	6.00	5.40
200.0	—	7.95	7.20	6.55	6.05

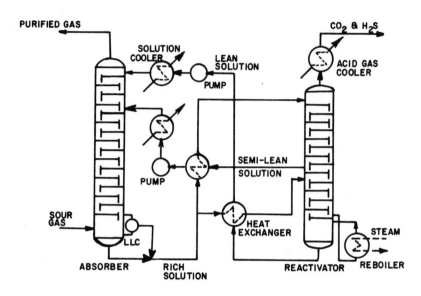

FIG. 7.11 The bubble column, or Girbotol, process.

CO_2 which vary with their concentration in solution, temperature, and partial pressure of the CO_2 in the raw gas. Compared to the other ethanolamines, MEA has a higher capacity per unit weight of solvent (although the difference is not great), a higher reactivity, and is more easily regenerated. MEA also has disadvantages which set it apart from the other ethanolamines, particularly in the presence of carbonyl sulfide, where a heat-stable compound is formed, resulting in loss of amine.

Sodium carbonate, potassium carbonate, and potassium phosphate have been used in natural gas scrubbing facilities as well. These processes have usually been applied to H_2S scrubbing; however, taking into account the different requirements inherent in the high CO_2 and water vapor characteristics of the raw digester gas, these technologies should be applicable.

Water vapor may be removed from the raw digester gas either before or in conjunction with the CO_2 scrubbing process. Depending on the specific approach, preremoval may raise the temperature of the raw gas, resulting in an increase in the energy required for CO_2 removal. Postabsorption drying will tend to cause dilution of the CO_2 scrubber working fluid, as water is condensed out of the gas during this process.

Another gas purification method which allows for simultaneous CO_2 removal and drying utilizes molecular sieve adsorbents. In this process, polar adsorbents such as synthetic zeolites (sodium or calcium aluminosilicates), ammonia, acetylene, hydrogen sulfide, and sulfur dioxide, are used. After adsorption, zeolite is heated and the polar molecules driven off.

A molecular sieve arrangement would involve the use of two or more containers. While one container is absorbing water vapor and CO_2, the other unit may be heated to regenerate the contents. As the contents of the containers approach their saturation levels, they can be taken out of stream and heated for regeneration while one of the other vessels can be put back on line to adsorb.

Caustic Scrubbing

For industrial gases containing large concentrations of CO_2 and H_2S, caustic scrubbing is done using either sodium hydroxide (NaOH), potassium hydroxide (KOH), or calcium hydroxide $[Ca(OH)_2]$. When aqueous solutions of these materials contact CO_2, successive reactions occur.

$$2NaOH + CO_2 \rightarrow Na_2CO_3 + H_2O \qquad (7.4)$$

$$Na_2CO_3 + CO_2 + H_2O \rightleftharpoons 2NaHCO_3 \qquad (7.5)$$

As illustrated in Eqs. (7.4) and (7.5), first an irreversible carbonate-forming reaction occurs and then a reversible bicarbonate-forming reaction.

The calcium hydroxide (or lime-water) scrubbing technique is widely used. Disadvantages in using this approach are the difficulties in controlling solution strengths and the removal of large quantities of $CaCO_3$ (the precipitate) from the scrubber and mixing vessel. Excessive precipitate will plug pumps, high pressure spray nozzles, and scrubber tower packing.

If sufficient contact time is allowed, H_2S can also be removed via caustic scrubbing, according to the following reaction:

$$H_2S + Na_2CO_3 \rightarrow NaHS + NaHCO_3 \qquad (7.6)$$

As shown, H_2S is precipitated by the reaction with the carbonate formed in the reaction described by Eq. (7.4).

In general, caustic scrubbing is rarely practiced in small-scale biogas systems.

8

Methanol Production
from Biomass

OVERVIEW OF ALCOHOL FUEL TECHNOLOGY

Alcohols are basically any class of chemical compounds that can be expressed by the general formula ROH; here R represents an alkyl group. Alcohols are derived from hydrocarbons by replacement of a hydrogen atom by a hydroxyl radical (OH).

Alcohols are classified as primary if there are no less than two hydrogen atoms attached to the carbon atom with the OH group. They are referred to as secondary if no hydrogen atoms are present. Primary alcohols with a straight-chain carbon structure are called normal (n). The prefix "iso-" refers to alcohols containing the terminal groups $(CH_3)_2 CH$ and no other branched chains. Figure 8.1 shows the chemical structure of some of the more commonly known alcohols used as fuels.

Alcohols can be derived from a number of sources and by several conversion techniques. Methanol is primarily produced from natural gas at present. Other alcohols are derived mainly from cracked petroleum feedstocks. A small portion of industrial ethanol is manufactured from various foods such as grain and molasses via fermentation and distillation. Figure 8.2 illustrates the various routes by which alcohols can be produced.

Methanol, also called methyl alcohol or wood alcohol, was recognized for its commercial fuel value during the pre-World War II era. At present there is only limited use of methanol as a fuel. Roughly 50% of the current world production of methanol is converted to formaldehyde, with the remainder being consumed in numerous other products.

Prior to the 1920s methanol was exclusively manufactured from the distillation of wood; today, as already noted, it is nearly all produced from natural gas. In order to reduce petroleum consumption, one proposal has been to blend methanol in gasoline for automotive use (153). The idea is not new. For example, alcohols were used in vehicles in the earlier part

158

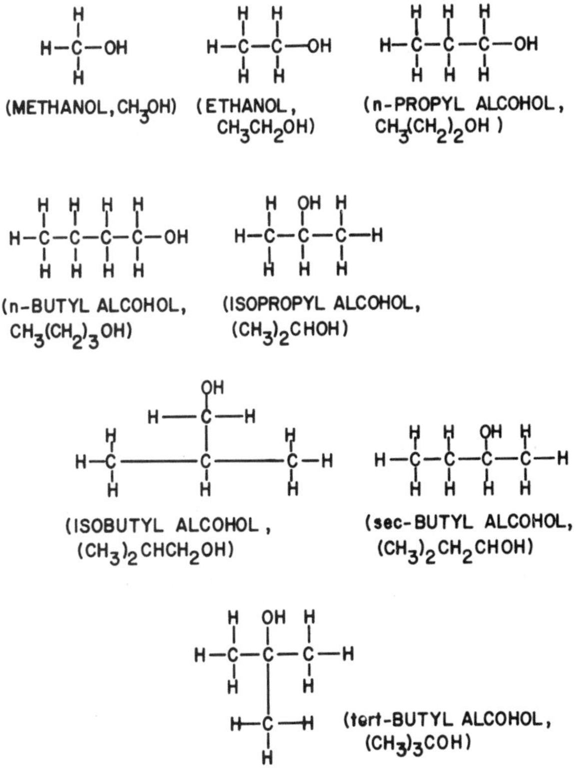

FIG. 8.1 Chemical structure of various alcohols.

of this century. Today special gasoline mixtures for racing cars use methanol and/or ethanol as a fuel. With the depletion of natural gas reserves in the Western Hemisphere, renewed interest has risen in the manufacture of methanol from other sources. Synthetic methanol from coal is one promising approach in addition to the technology discussed next.

PRODUCTION OF METHANOL

There are two primary reactions that can be used to generate methanol (CH_3OH). These reactions are the synthesis gas feedstock reaction [Eqs. (8.1) and (8.2)] and hydrocarbon oxidation [Eq. (8.3)].

$$CO_2 + 3H_2 \rightarrow CH_3OH + H_2O \qquad (8.1)$$

$$CO + 2H_2 \rightarrow CH_3OH \qquad (8.2)$$

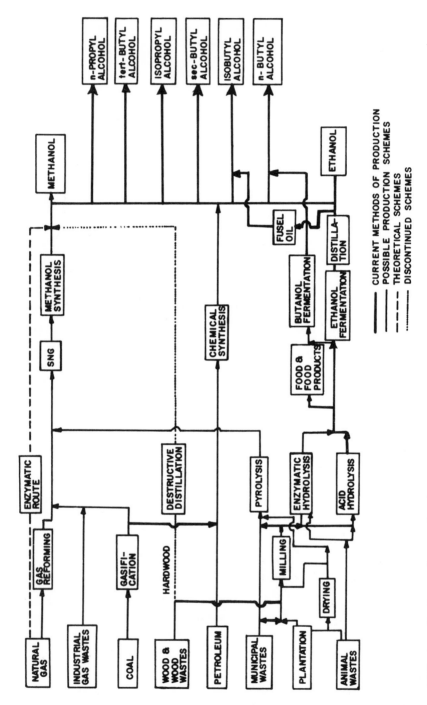

FIG. 8.2 Various schemes for the manufacture of alcohols.

In the foregoing reactions, the mixture of CO, CO_2, and H_2 is called syn-thesis gas (SNG).*

In the hydrocarbon oxidation reaction, the following occurs:

$$2CH_4 + O_2 \rightarrow 2CH_3OH \tag{8.3}$$
(methane)

Other hydrocarbons in addition to methane can be used in methanol production.

Synthesis gas can be prepared from carbonaceous sources as well (e.g., coal). The basic reactions which occur are

$$C + H_2O \rightarrow CO + H_2 \tag{8.4}$$

$$C + 2H_2O \rightarrow CO_2 + 2H_2 \tag{8.5}$$

$$C + 2H_2 \rightarrow CH_4 \tag{8.6}$$

As we see from Eq. (8.6), methane can also be formed from carbonaceous sources.

Any light hydrocarbon such as methane can be used as a source for SNG. Examine the following reactions.

$$CH_4 + H_2O \rightarrow CO + 3H_2 \tag{8.7}$$

$$CH_4 + 2H_2O \rightarrow CO_2 + 4H_2 \tag{8.8}$$

$$CO + H_2O \rightarrow CO_2 + H_2 \tag{8.9}$$

---

*Synthesis gas should not be confused with the term syngas (i.e., synthetic gasoline), used elsewhere in this chapter.

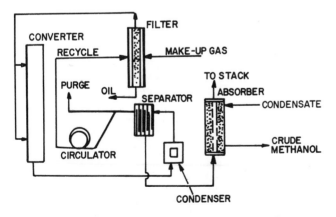

FIG. 8.3 Flow diagram for methanol production from SNG.

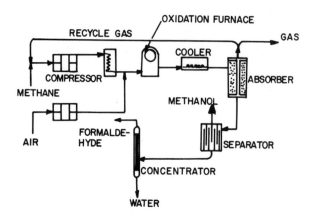

FIG. 8.4   Flow diagram for methanol production from partial oxidation of methane.

The SNG containing hydrogen and carbon monoxide or carbon dioxide is passed over a catalyst under pressure and at high temperatures. Catalysts normally used are chromium, zinc, or copper oxides with pressures and temperatures ranging from 50 to 350 atm and up to 400°C, depending on the catalysts and SNG mixture (154). SNG can be obtained from the gasification of fossil fuels. The gas is purged of all sulfur compounds in order to protect the catalyst. Figure 8.3 illustrates the process by which methanol is produced from SNG.

Methanol production from partial oxidation of methane is illustrated in Fig. 8.4. In general, any hydrocarbon can be partially oxidized to produce some methanol; however, yield will decrease as the average carbon number of the feedstock increases in the absence of a highly selective catalyst.

METHANOL FROM BIOMASS

Earlier, coal was cited as an example of carbonaceous material that could be used for synthetic methanol production. Other examples are lignite, wood waste, agricultural residue, and urban solid wastes. These feedstocks, however, require additional processing steps to refine crude gas product into a final syngas, in contrast to liquefied natural gas (LNG).

In general, the conversion of a carbonaceous material to syngas is significantly more energy intensive than that required to liquefy natural gas. Furthermore, yields are generally less.

To convert solid carbonaceous material to a syngas, it is necessary to partially oxidize the material to form a crude gas consisting primarily of H_2, CO, and CO_2. Air can be used to oxidize the feed material, resulting

in a crude gas containing about 46% nitrogen, which can be removed cryo-
genically. If oxygen is used rather than air, a cryogenic system is necessary for initial separation of air into oxygen and nitrogen.

Several types of gasifiers have been developed for the partial oxidation
of wood, wood residue, and garbage. They have been designed to operate
at atmospheric pressure, in contrast to coal gasifiers which can operate
at pressures of up to 27 atm. The gaseous product from a gasifier consists
primarily of H_2, CO, and CO_2, with minor amounts of hydrocarbons.
Roughly 2% of the wood (dry basis) is converted to an oil-tar fraction.

The basic steps for production of methanol from wood residues are outlined below:

1. The wood residue is partially oxidized.
2. Crude gas is cleaned and cooled.
3. The gas generated is compressed to about 7 atm.
4. Carbon dioxide is removed.
5. Residual carbon dioxide is also removed.
6. Removal of nitrogen and hydrocarbons takes place.
7. The gas is compressed to about 27 atm.
8. Through the use of a catalyst, the stoichiometric ratio is shifted to
    two parts hydrogen to one part carbon monoxide.
9. Removal of carbon dioxide formed during the stoichiometric shift
    is accomplished.
10. The gas is then compressed to about 170 atm.
11. Hydrogen and carbon monoxide are converted to methanol.
12. The final stage involves the refinement of crude methanol.

Moore-Canada of Richmond, British Columbia, has developed a moving
bed reactor for producing low-Btu gas from wood biomass. The reactor
uses air as the oxidizing medium. Due to the high nitrogen content, the
product gas has a heating value of around 180 Btu/SCF. In this system,
the feedstock enters through the top of the reactor and wood ash is discharged
from the bottom. Because of the use of air as an oxidizing medium, the
maximum temperature of the oxidation zone is about 2200° F and consequently
product wastes are discharged in solid, granular form rather than as a
molten slag. Crude gas is discharged from the reactor at about 160-180° F
(155).

Municipal wastes are another source of biomass feedstock for methanol
production. Several commercial processes are being utilized. Some of
these systems burn refuse directly for conversion to heat for electric power
generation. Other systems combust a combination feedstock of coal and
refuse. Still other systems pyrolyze wastes to gaseous products that include methane, hydrogen, carbon monoxide, carbon dioxide, and water.
These products can either be used directly for their heating value or be
converted to methanol as previously described.

Union Carbide has developed a process for the partial oxidation of municipal trash without any feedstock preparation. The system, called Purox, utilizes oxygen for the oxidation step (156). The reactor-gasifier is similar to the Moore reactor in that it uses a moving bed. Oxygen is passed countercurrently over the downflowing residue. As the biomass travels through the reactor, it passes through successive stages of drying, reduction, and oxidation. Ash is removed at the bottom of the reactor in a molten state at temperatures up to 3000° F. Crude gas is discharged from the top of the reactor at about 200° F.

USE OF METHANOL AS A FUEL

Automotive Uses

Methanol has been suggested as an additive to gasoline. Theoretically, 14.5 parts of air to one part of gasoline is required to obtain the proper combination for complete combustion. For methanol the air-to-fuel ratio is roughly 6.4. The addition of methanol to gasoline causes automatic chemical leaning in a car, that is, a lower air-to-fuel ratio is automatically reached even without altering the engine's carburetor settings. Engines run on such a mixture generally experience fewer emissions. Unfortunately, this is often accompanied by a drop in performance. The Btu capacity of gasoline is roughly twice that of methanol; that is, roughly twice the amount of methanol is necessary to drive the same distance as with gasoline.

A review of the literature shows that there is disagreement among various researchers as to the effects of methanol-gasoline mixtures. Reed, Lerner, and their associates have tested unmodified cars over a fixed course using mixtures of methanol and gasoline (153, 157, 158). They observed that mixtures between 5 and 15% increased fuel economy and performance and lowered carbon monoxide emissions and exhaust temperatures. These investigators attributed the improved methanol performance to chemical leaning and to the dissociation of methanol near 200° C. At this temperature, energy can be absorbed during the compression stroke of the engine and release up to 40% hydrogen to give a 10% mixture (159). Other such studies are cited in Refs. 160-166.

Experience with 100% methanol as a fuel in automobile engines has been limited to racing cars. It is known that methanol tends to develop more power than gasoline and that compression ratios as high as 15:1 are possible with methanol fuel.

Starkman et al. (167) made performance comparisons of methanol, ethanol, benzene, and isooctane as tested on a CFR test engine. In such testing, isooctane is often used as a pure chemical representative of gasoline. At stoichiometric equivalence ratios, methanol was found to develop slightly more power than isooctane. At very rich equivalence ratios (values around 1.5:1) methanol was observed to develop significantly more power than isooctane. The study demonstrated that, with this increased power, methanol shows a significantly greater fuel consumption than isooctane.

Electric Power Generation

The use of methanol has also been suggested as a fuel for electric power generation. In one study, a commercial utility boiler having 50-MW capacity was used for a large-scale demonstration. Test data showed that methanol not only provided a good stable flame and good burning efficiency but also better performance than natural gas or fuel oil in reducing pollutant nitrogen oxides. Methanol was found to combust satisfactorily in existing utility boilers originally designed for alternate firing with or without fuel oil (168). The advantages of using methanol for electric power generation include: (a) no particulates; (b) lower nitrogen oxides than from natural gas combustion; (c) lower carbon monoxide concentrations than from oil and gas firing; and (d) no sulfur compounds emitted.

In this study, tests for emission of aldehydes, acids, and unburned hydrocarbons indicated that insignificant quantities were emitted during methanol combustion.

General Electric Company performed a series of tests using methanol as a fuel for gas turbines (169). As a base case fuel, No. 2 distillate was used. Three types of methanol fuel were tested: 100% dry methanol; an 80% methanol/20% water mixture; and 75% distillate/25% methanol blend. To compensate for the Btu differences in all of these fuels, the actual mass flows of the fuel into the turbine were varied accordingly.

General Electric reported that some retrofitting is required before turbine generators can run on methanol fuel, because of the low lubricity of methanol. Pumps built to transfer fuel oil use the fuel itself for their own lubrication. Effective lubrication would not be achieved with methanol.

In addition to the large-scale concepts of methanol as a fuel for fixed electric power generators, interest has also been generated in its use as a fuel for fuel cells. Such fuel cells fall into two categories: (a) those using the methanol directly as the fuel in the electrolyte, and (b) those using the methanol as a source of hydrogen. In the latter, methanol is converted to hydrogen by steam reforming.

Most fuel cells employed by the U.S. military use hydrazine as a fuel, which is highly toxic. Thus, methanol would be much more desirable than hydrazine as fuel for general public use.

Natural Gas Industry Usage

Natural gas can be liquefied by the application of pressure and very low temperatures and then be shipped in liquid form (LNG) to the consumer. In order to reduce the high costs associated in LNG storage and tanker facilities for shipment over long distances, suggestions have been made that natural gas be converted to methanol and shipped in ordinary tankers.

The methanol can then be reconverted to methane for use in the natural gas system. The advantages of such an approach as outlined by McGhee (170) are the following:

1. Methanol can be shipped in conventional tankers used for hauling crude oil or refined products. LNG requires much more expensive, specially designed single-purpose tankers.
2. Methanol can be shipped to any port and offloaded into conventional terminal tankage used for crude oil, refined products, petrochemicals, etc.
3. Methanol can be delivered from such a terminal through existing distribution facilities, such as pipelines, rail cars, or tank trucks.
4. Methanol can be stored inexpensively for long periods of time for peak load or seasonal demand uses.

The vapor pressure and density of methanol are similar to those of gasoline, and so transportation and storage of methanol should present no more problems than gasoline does. Methanol, like other alcohols, can also be transported through steel pipelines.

One serious problem related to the storage and transportation of methanol destined for fuel use is the affinity of alcohol for water. Substantial quantities of water are present in gasoline vessels, especially in storage tanks, ships, and barges. Since gasoline and water are not miscible, this now poses no problems. Water forms a layer at the bottom of a gasoline tank and is usually undisturbed. Moreover, this layer of water is actually beneficial in that it catches any sediment from the gasoline and also is a good barrier between the gasoline and any bottom leaks in the tanks. If a leak should develop in the bottom of a gasoline tank, the water leaks out first and discovery of the leak is possible before actual loss of gasoline. With methanol, however, this water barrier would not form. Any water present would be drawn into solution with the alcohol.

If alcohol is blended with gasoline, then the amount of water in a tank becomes critical because of phase-separation problems associated with water contamination. Such water-prevention systems would add a large expense to any transportation and storage system.

If methanol is to be used separately from gasoline, then small quantities of water might not present serious problems. However, were contamination to occur with seawater, increased corrosion problems would probably result from alcohol used as fuel.

If the use of methanol as a fuel becomes large scale, one factor that must be accounted for in cost analyses is its reduced Btu content in comparison to that of hydrocarbon fuels. With methanol, twice the quantity on a volume basis will be necessary to fulfill the same requirements as gasoline or fuel oil. Obviously, this would necessitate increased shipping and storage capacity in the transportation network. Johnson (171) has performed a preliminary study of the transportation and storage requirements of synthetic fuels, including methanol.

Use as a Feedstock for Food

Methanol is toxic to most animal and plant life. There are, however, a number of species of single-cell algae, bacteria, and yeasts that will grow well on a feedstock of methanol and inorganic nutrients. These species have a notably high protein content, and some varieties are presently being substituted for milk or soybeans in calf feeds. Thus, methanol might be considered to be an indirect feedstock in the manufacture (i.e., metabolism) of single-cell protein.

Hagen (154) notes that the production of single-cell protein could have a significant impact on the world's food supply. He points out that large-scale energy plants in arid regions throughout the world could conceivably use atmospheric carbon dioxide and water to produce methanol. Methanol produced could, in turn, be used as feedstock for production of single-cell protein.

9

Ethanol and Other
Wood Alcohols

ETHANOL SYNTHESIS

Ethanol, or ethyl alcohol, is the well-known alcohol suitable for human consumption. It can be manufactured by two processes: (a) direct hydration of ethylene gas, or (b) fermentation and distillation using food feedstocks. There is basically no chemical difference between ethanol produced from either process; however, the alcohol manufactured for consumption is required by law to be made by the latter method.

Ethanol production through the fermentation of grains and fruits has been practiced for centuries. The process of fermentation is basically a decomposition reaction involving interaction between various enzymes and sugar solutions. Enzymes serve as catalysts, as is illustrated by the fermentation of sucrose:

$$C_{12}H_{22}OH + H_2O \xrightarrow{\text{invertase}} 2C_6H_{12}O_6 \tag{9.1}$$
(sucrose)

$$C_6H_{12}O_6 \xrightarrow{\text{zymase}} 2C_2H_5OH + 2CO_2 \tag{9.2}$$
(invert sugar)

Feedstocks for these reactions can consist of almost all forms of plant biomass. Important examples include sugar beets, potatoes, corn, sugar, molasses, wheat, and various other grains. Recall that plant matter contains cellulose, which can be hydrolyzed into glucose. Residues from the paper and lumbering industries are presently converted to glucose via acid hydrolysis.

The actual fermentation process takes approximately 38-48 hr. The liquid product generally contains about 6-12% ethanol, which can be distilled off. During distillation, various impurities including aldehydes and esters

TABLE 9.1 Various Properties of Ethanol and Isooctane

Property	Ethanol (C_2H_5OH)	Isooctane (C_8H_{18})
Octane number	106	100
Molecular weight	46.07	114.224
Weight percent carbon	52.0	84.0
Weight percent hydrogen	13.0	16.0
Weight percent oxygen	35.0	0.0
Boiling point (°F)	173.0	210.6
Freezing point (°F)	-173.4	-161.3
Specific gravity (60°F/60°F)	6.63	5.795
Specific heat at 77°F, 1 atm (Btu/lb-°F)	0.6	0.5
Heat of vaporization at 77°F, 1 atm (Btu/lb)	395	132
Heat of combustion at 77°F (Btu/lb):		
Higher heating value	12,780	20,556
Lower heating value, liquid fuel—gaseous H_2O	11,550	19,065
Stoichiometric mixture (lb air/lb)	9.0	15.3
Autoignition temperature (°F)	685	784

are distilled off along with the ethanol. Commercial grades of alcohol generally contain roughly 95% ethanol, which is 190 proof. Continued distillation does not increase purity as the composition is a constant boiling point mixture of water and alcohol.

At present, the majority of ethanol production is carried out by direct hydration of ethylene via catalysis. The reaction, described by Eq. (9.3), is carried out at about 1000 psi and 400°C with phosphoric acid as the catalyst:

$$C_2H_4 + H_2O \xrightarrow{H_3PO_4} CH_3CH_2OH \tag{9.3}$$

(Note that there is another process, known as the indirect hydration process, which can be used for ethanol production. However, this method is no longer widely used.)

Ethanol is probably the second most widely used alcohol in automobile engines. Table 9.1 compares some of the properties of ethanol to isooctane. In various parts of the world, ethanol is mixed with gasoline for automobile engines. In Brazil, for example, ethanol production helps to support the price of the sugar crop. Excess sugar crops are purchased by the Brazilian government and fermented and distilled to ethanol, which is then blended with gasoline and sold to the public as "gasohol."

The use of ethanol in gasoline in the United States is technologically feasible; however, it has not been considered economically competitive with petroleum products. However, this situation is likely to change soon if the price of petroleum continues to rise sharply and the production of ethanol from biomass sources such as agricultural and municipal wastes is developed and expanded.

In comparison to methanol, ethanol has a significantly higher heating value (roughly 12,780 Btu/lb for ethanol and 9500 Btu/lb for methanol). Both values, however, are significantly lower than that of gasoline. To illustrate, a gallon of ethanol contains roughly 0.7 the Btu capacity of gasoline. Consequently, the addition of ethanol to gasoline results in a drop in the Btu capacity. The addition of either methanol or ethanol to gasoline causes an automatic leaning of the fuel mixtures due to the difference in stoichiometry of the two fuels.

Ethanol also has a high octane rating (106-107.5 RON and 89-100 MON). The addition of ethanol to nonleaded gasoline causes the octane rating to increase as well as the antiknock capacity of the fuel.

BIOCONVERSION TECHNIQUES

Since ethanol can be produced from cellulose, potential feedstocks include refuse, sewage, and animal and agricultural residues. Ethanol synthesis from such feedstocks can be accomplished by the hydrolysis of cellulose to glucose, followed by fermentation and distillation. Hydrolysis experiments have been reported by Porteous (172), Converse et al. (173), Grethlein (174), and Chapman (175).

Acid hydrolysis methods are commercially employed for the conversion of wood wastes into glucose. In this technique, weak acid solutions are combined with wood chips in a digester. The mixture is cooked to produce glucose.

In the enzymatic hydrolysis process, cellulose is transformed into glucose through the use of enzymes. A cellulose culture can be grown and added to a digester containing milled cellulose. After saccharification, crude glucose syrup is removed from the batch product by filtration and then reacted cellulose and enzymes are recycled back to the reactor (176,177).

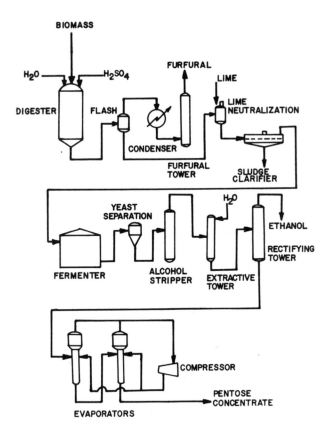

FIG. 9.1 Basic process for ethanol production via sulfuric acid hydrolysis
and fermentation.

This process has been found to have several operational problems. For
one, lignin is a deterrent to the hydrolysis process in that it slows down
the reaction. Also, the rate of enzymatic hydrolysis is very sensitive to
the form of cellulose in the substrate. In order to achieve adequate decomposition of the cellulose, it is necessary that the substrate be relatively
clean and well milled. It should be noted that such milling operations add
considerable expense to the system.

The most advanced process in terms of operating experience is the
Scholler process (83). Figure 9.1 schematically shows the basic process.
Biomass is discharged from a hopper on the front-end unit into a digester.

Steam is introduced to raise the feedstock temperature to roughly 257° F. Upon reaching this temperature, sulfuric acid and a recycled dilute, previously hydrolyzed stream are introduced into the digester, followed by the addition of hot water. The slurry is then maintained at a constant temperature of 300° F for a period of about 15 min (83).

Prehydrolyzate is drained, and dilute sulfuric acid is introduced from the top of the digester along with high pressure steam. High pressure steam, at about 250 psig, raises the temperature to about 300° F, which provides the necessary conditions for the main hydrolysis. Digestion takes place for almost 1.25 hr, at the end of which the main hydrolyzate is flashed to the atmosphere in stages. A stage-flashing operation also serves to separate out the lignins, fufural, and methanol.

Collected lignins are returned and used as fuel. Methanol and fufural are sent to a distillation stage for methanol recovery. The hydrolyzate is neutralized in a lime slurry vessel, and precipitated calcium sulfate is separated in a clarifier. The precipitate is then washed to recover sugar. Sugar solutions collected are mixed and sent to the fermentation tanks for alcohol production. At the end of the fermentation process, yeast is removed and the crude alcohol-water mixture sent to distillation to separate out pentoses and concentrate the ethanol. The process generates a considerable amount of sludge, consisting primarily of calcium sulfate, which requires daily removal.

SYNTHESIS OF PROPYL AND BUTYL ALCOHOLS

Propyl

Isopropyl alcohol and n-propyl alcohol are the two monohydric propyl alcohols. The latter is made from ethylene gas via oxo technology or by an oxidation process from propane. N-propyl alcohol can also be produced in small quantities as a by-product in the distillation of fermented products. N-propyl alcohol, like most alcohols, is primarily used as a solvent and as a chemical intermediate.

Isopropyl alcohol is manufactured via hydration of propylene. As with ethanol production, the final distillation of this alcohol produces a binary mixture of isopropyl alcohol (roughly 91.3% by volume of isopropyl) and water. The alcohol is either sold as a water solution or further dehydrated with the use of an azeotropic agent such as benzene and isopropyl alcohol. Figure 9.2 illustrates some of the uses of isopropyl alcohol.

Butyl

Four monohydric butyl alcohols that have been considered as fuel alternatives are n-butyl alcohol, isobutyl alcohol, sec-butyl alcohol, and tert-butyl alcohol.

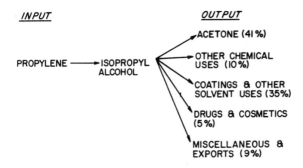

OUTPUT

PROPYLENE ──────► ISOPROPYL
                  ALCOHOL

ACETONE (41%)

OTHER CHEMICAL
USES (10%)

COATINGS & OTHER
SOLVENT USES (35%)

DRUGS & COSMETICS
(5%)

MISCELLANEOUS &
EXPORTS (9%)

FIG. 9.2 The uses of isopropyl alcohol.

There are five production methods for the synthesis of n-butyl alcohol:

1. Fermentation
2. The aldol process
3. The oxo process
4. High-pressure oxidation of propane and butane
5. The Ziegler process

In the fermentation process, a carbohydrate mash is acted on by a bacterial culture, namely Clostridium saccharobutyl acetonicum liquefaciens. Normally a 36- to 48-hr period is required for fermentation, after which distillation produces 30% (by weight) of the theoretical yield of 60% n-butyl alcohol, 30% acetone, and 10% ethanol (166).

In the aldol process, the feedstock consists of either ethyl alcohol or acetylene. The intermediate chemical that appears in the process is crotonaldehyde. Approximately 85% of the theoretical yield is achieved.

The majority of producers of n-butyl alcohol use the oxo process. In this process propylene is used as the feedstock.

The Ziegler process (developed by the Continental Oil Company) was originally developed commercially for the manufacture of even-numbered carbon alcohols. N-butyl alcohol as a C_4 alcohol is at the low end of this production line and as such is a by-product.

Isobutyl alcohol is produced mainly as a by-product from high-pressure methanol synthesis. The Fischer-Tropsch oxidation process for acetaldehyde production is another source of isobutyl alcohol. In this process, butane or propane can be used as a feedstock. The alcohol is also a major constituent in molasses, comprising roughly 74% of the alcohols present in fusel oil resulting from fermentation. Isobutyl alcohol also represents between 12% and 24% of the alcohols obtained from the fermentation of corn.

The production of sec-butyl alcohol is carried out by absorbing butylene on sulfuric acid. This forms butyl hydrogen sulfate, which is then reacted with steam. The principal use of sec-butyl alcohol is as a chemical inter-

mediate in the production of the solvent methyl ethyl ketone; it also has use
as a solvent and a coupling agent in hydraulic brake fluids, cleaning compounds, and paint removers.
      tert-Butyl alcohol is manufactured by a process similar to that used for
sec-butyl alcohol. In this case, however, isobutylene is employed as the
feedstock. The isobutylene is absorbed by sulfuric acid to form isobutyl
sulfate, which is then steamed to produce the tert-butyl form. In general,
this alcohol is very miscible in a large number of organic compounds,
making it useful as a blending agent. Specific uses of tert-butyl alcohol
are (a) as a selective solvent and extraction agent in the pharmaceutical
industry; (b) in the removal of water from compounds; (c) in the manufacture
of perfumes; and (d) in the purification of chemicals via recrystallization.

FUEL USES OF ALCOHOLS

As already noted, ethanol as a fuel source for the internal combustion
engine is not a new idea. In Europe, during World War II, ethanol-gasoline
mixtures were common. In many parts of the world today, ethanol is mixed
with the gasoline supply.
      In this country, considerable interest has developed in the large-scale
production and use of ethanol as an automotive fuel only in recent years.
In the early 1970s, Miller (178-180) indicated that the use of ethanol in
gasoline, while technologically feasible, was not yet competitive economically with petroleum products. However, as we noted earlier, the ongoing
energy crisis is profoundly changing the global economic picture.
      The problems associated with the use of ethanol-gasoline mixtures are
similar to those experienced with methanol-gasoline blends. Some of these
problems are in the areas of starting, performance, and phase separation
of ethanol and gasoline because of water contamination.
      Propyl and butyl alcohols have had limited use as fuels. Examples
include n-propyl alcohol, employed as an antifungicide in jet fuel, and
isopropyl alcohol, used in high-compression aircraft engines as an anti-
detonation injection fluid with water and as a deicer for automobile fuel.
      tert-Butyl alcohol has had wider use as a blending component in auto-
mobile fuels than the other butyl and propyl alcohols discussed. tert-Butyl
alcohol-gasoline blends are capable of accepting much more water in solution before phase separation than both methanol- and ethanol-gasoline
mixtures.
      Fuel properties can be compared by examining data on heat content,
ignition temperature, etc., for individual substances in a blend. Table
9.2 contains such information on the fuel properties of the various alcohols
discussed in this section.

TABLE 9.2 Properties of Various Alcohols

Alcohol	Common Name	Chemical Formula	Molecular Weight	Octane Number (MON)	Normal Boiling Point (°F)	Freezing Point (°F)	Heat of Vaporization (cal/g)	Heat of Combustion (cal/g)
Methyl alcohol	Methanol	CH_3OH	32.042	88–92	32	-144	262.8	5334
Ethyl alcohol	Ethanol	C_2H_5OH	46.070	89–100	173	175	204.3	7120
Propanol-1	n-Propyl alcohol	$CH_3CH_2CH_2OH$	60.097	–	207	-195	162.6	8000
Propanol-2	Isopropyl alcohol	$CH_3CHOHCH_3$	60.097	92.8–98.5	189	-127	165.0	7970
	n-Butyl alcohol	$C_2H_5CH_2CH_2OH$	74.124	81.5–85	255	-129	141.3	8626
Butanol-2	sec-Butyl alcohol	$C_2H_5CHOHCH_3$	74.124	–	210	-174	134.3	–
	Isobutyl alcohol	$(CH_3)_2CHCH_2OH$	74.124	–	224	-162	137.2	8610
	tert-Butyl alcohol	$(CH_3)_3COH$	74.124	–	181	78	130.4	8490

ECONOMIC CONSIDERATIONS

There are a number of factors that will influence the costs of biomass conversion processes for the production of alcohol fuels. Some of these factors
are regional variations, resource availability, demand profiles, contractor
markups, fees, and contingency charges, construction time, and effects of
inflation.

Regional variations may be expected to play an important role in the cost
of process equipment. Labor and material costs vary considerably from
region to region. In addition, energy prices vary markedly in different
parts of the country.

A major limitation on the use of such biomass processing plants nationwide is resource availability. This may take the form of high costs for
materials and components for plant construction, or inadequate raw materials
for the process itself. Difficulty may be experienced in the installation of
large process assemblies which have previously only been used in pilot
plants; or difficulty may be experienced in obtaining adequate and consistent
supplies of raw materials, resulting in lengthy production delays.

A major element in the costing of processes is the determination of lead
time for response to a purchase order. Large-scale systems (for example,
utility-owned systems) can experience lead times ranging from 7 to 13 years
(83). These figures are typical for large-scale power plant installations
such as a coal-fired plant having some 1000-MW capacity. It is likely that
biomass conversion systems will be considerably smaller and, hence, lead
times might be on the order of 2-4 years.

Construction times necessary for biomass systems are expected to be
relatively short. Many of the processes discussed are fairly simple. An
exception to this is the acid hydrolysis system for the production of ethanol
which is a relatively complicated process requiring considerable plant construction.

Case Study

The Mitre Corp. (83) performed an economic study based on three
different-sized plants for several bioconversion systems. The three plant
sizes evaluated were based on a biomass feedstock input of 850 ODT (oven
dry ton)/day, 1700 ODT/day, and 3400 ODT/day. The smallest plant capacity is suited to a biomass tree farm of approximately 250,000 ODT/year,
whereas the two larger facilities would probably require the output from
several biomass farms.

In this study, silvicultural biomass was considered to be the only feedstock and a standardized front-end process was selected for evaluation.
The materials-handling design for the systems studied is illustrated in
Figure 9.3. It was assumed that the feedstock is to be received at the plant
site on a two-shift, 16-hr day basis from trucks with capacities of 40,000 lb.

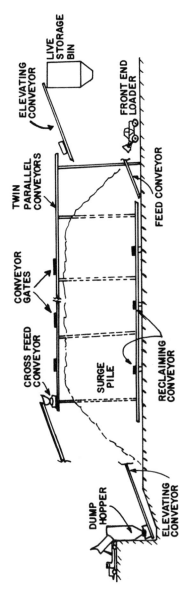

FIG. 9.3 Front-end handling and inventory design schematized by Blake and Salo (83) in the Mitre Corp. study.

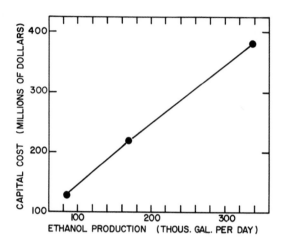

FIG. 9.4 Capital cost estimation based on 1977 dollars for ethanol production as a function of plant size. [Data obtained from Balke and Salo (83).]

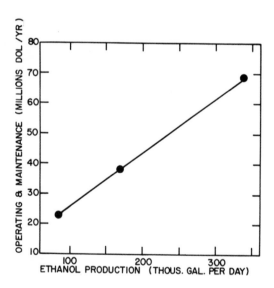

FIG. 9.5 Operating and maintenance (O&M) costs in 1977 dollars for
ethanol production as a function of plant size. [Data obtained from Blake
and Salo (83).]

TABLE 9.3 List of Items Considered in Capital and Operating Costs

	Plant Size		
Feed capacity, ODT/day	850	1,700	3,400
Ethanol production, gal/day at 190 proof	84,400	168,800	337,600

Capital Cost Items

Front-end process	Evaporation and concentration equipment
Hydrolysis equipment	Steam plant
Crude furfural and ethanol equipment	Electric generation
Fermentation equipment	Water treatment plant
Cooling towers	Building and structures
Storage and shipping	Site development
Fire protection	Engineering and fees at 15%
Instrumentation and controls	Commissioning and contingency at 15%
General mechanical equipment	Working capital at 10% (or more)
General electrical equipment	

Operating and Maintenance Costs

Direct labor cost	Utility costs
Labor-related cost	Maintenance costs
Power costs	Equipment repair costs
Chemical costs	

TABLE 9.4 Physiological Effects of Methanol and Ethanol on Animals

Animal	Concentration ppm	mg/liter	Duration of Exposure[a]	Signs of Intoxication	Outcome
Methanol[b]					
Cat	132,000	173.0	5-5.5	Narcosis	Died
	33,600	44.0	6	Incoordination	50% died
Mouse	72,600	95.0	54	Narcosis	Died
	54,000	70.7	54	Narcosis	Died
	48,000	62.8	24	Narcosis	Survived
	10,000	13.1	230	Ataxia	Survived
	152,800	200.0	94 min	Narcosis	
	101,600	133.0	91 min	Narcosis	
	91,700	120.0	95 min	Narcosis	Overall
	76,400	100.0	89 min	Narcosis	mortality
	61,100	80.0	134 min	Narcosis	45%
	173,000	227.0			Died
	139,000	182.0		Highest concentration endurable	
Rat	60,000	78.5	2.5	Narcosis, convulsions	
	31,600	41.4	18-20		Died
	13,000	17.0	24	Prostration	
	8,800	11.5	8	Lethargy	
	3,000	4.0	8	None	
	50,000	65.4	1	Drowsiness	Survived
Dog	37,000	48.4	8	Prostration, incoordination	
	13,700	17.9	4	None	
	2,000	2.6	24	None	
Monkey	40,000	52.4	4	Illness	Death
Rabbit	40,000	52.4	1 daily		Delayed death
Rat	10,000	13.1	18 daily		Death

(continued)

Table 9.4 (continued)

Animal	Concentration ppm	mg/liter	Duration of Exposure[a]	Signs of Intoxication	Outcome
Ethanol[c]					
Mouse	31,900	70.0	0.33	Ataxia	
	29,300	55.0	7.0	Narcosis	Died
	23,940	45.0	1.25	Narcosis	
Guinea	45,000	84.6	3.75	Incoordination	
Pig	50,170	94.3	10.2	Deep narcosis	Died
	20,000	37.6	6.5	Incoordination	
	21,900	41.2	9.8	Deep narcosis	Died
	13,300	25.0	24.0	Light narcosis	
	6,400	12.0	8.0	None	Survived
Rat	32,000	60.1	8.0		Some died
	16,000	30.1	8.0		Some died
	44,000	82.7	6.5	Deep narcosis	Died
	19,260	36.2	2.0	Light narcosis	
	18,200	34.2	1.0	Excitation	
	22,100	41.5	15.0	Deep narcosis	Died
	10,750	20.2	2.0	Incoordination	
	12,700	23.8	21.75	Deep narcosis	Died
	6,400	12.3	12.0	Light narcosis	Survived
	3,260	6.1	6.0	None	
	4,580	8.6	21.13	Ataxia	Survived

[a]Numbers represent hours unless otherwise indicated.
[b]TLV 200 ppm.
[c]TLV 1000 ppm.
Source: Data from Refs. 181 and 182.

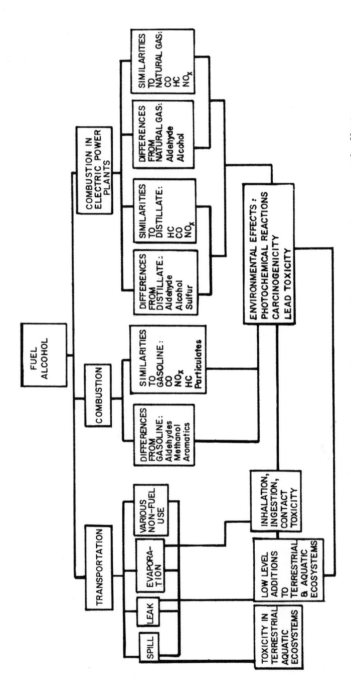

FIG. 9.6 Some of the proposed uses of alcohol as a fuel and their probable environmental effects.

The unloading hopper is designed to provide a 1-hr capacity for each of the three plant sizes, feeding a transfer conveyor which in turn feeds parallel overhead stockpile feeder conveyors equipped with distributing gates. Fixed-position reclaimer conveyors transfer stockpiled material to a conveyor that feeds the live-storage bin.

Total capital investment and O&M (operating and maintenance) costs for ethanol production via sulfuric acid hydrolysis and fermentation as a function of production rate are illustrated in Figs. 9.4 and 9.5, respectively. The individual components for which estimates were made for this economic study are listed in Table 9.3.

ENVIRONMENTAL AND PHYSIOLOGICAL
CONSIDERATIONS

It was noted in Chap. 8 that methanol as a fuel for the automobile appears to have environmental advantages in that NO_x emissions can be significantly reduced. For methanol-gasoline blends, the specific environmental effects are not yet clear. Ethanol-gasoline blends appear to show no improvement in reduction of emissions over straight gasoline. However, for electric power generation, improvement in air quality may be significant as no sulfur problems and very reduced NO_x emissions might be expected.

Aldehyde emissions have been noted to increase when using alcohols in automobiles. The effects of large-scale aldehyde dispersions have not yet been assessed.

The toxicology of alcohols under various controlled conditions is well-documented in the literature. It should be pointed out, however, that under uncontrolled situations, as would be the case with wide-scale use of alcohol-based fuels, the physiological and long-term environmental effects are not known. Vapor problems can certainly occur with evaporation of alcohols in poorly ventilated areas, and there may also be a fire hazard.

One subject that has not been investigated in detail is the environmental impact caused by spillage of alcohols. If large-scale shipping of alcohols were to become common, the effects on marine life might be serious should there be a major spill from an oceangoing tanker. As alcohols are almost completely soluble in water, cleanup of such a spill in the ocean or other waterway would be virtually impossible. The impact might well be disastrous, its severity depending on the type and quantity of alcohol spilled. Methanol, for example, is known to be extremely toxic to animals when taken internally. Some of the physiological effects of different alcohols on animals reported by various investigators are given in Table 9.4 (181, 182). Figure 9.6 summarizes some of the potential environmental effects to be considered when weighing the pros and cons of large-scale alcohol usage as fuel.

10

Energy Supplies from Sea Farming

TECHNOLOGY OVERVIEW

The basic advantage of using solar energy through biomass conversion is that plant life can serve as a continuously renewable energy collection and storage network. As already noted, through photosynthesis, plants can fix atmospheric carbon in the presence of sunlight, water, and certain inorganics. Resultant plant tissue can then be transformed into energy or energy-sparing substances through further biological and/or chemical processing. Some major considerations in the selection of biomass substrate are the following:

1. Its seasonality of production
2. Its application to other uses (for example, animal feed)
3. Its location with respect to the conversion facility

These considerations limit the large-scale applicability of many biomass sources. Problems associated with local availability and continuity of supply might be overcome through the use of energy farm systems. As noted for agricultural and silvicultural systems, areas can be selected to provide high-yield crops. However, long growing seasons are a prerequisite in order to supply feedstock continuously to the conversion facility.

The major difficulty in implementing large-scale bioconversion techniques is the low efficiency of the photosynthetic process. We noted earlier that the maximum efficiency of this process is only on the order of 8%; however, in practice, the actual efficiency is much lower (2-3%). This imposes a severe limitation on the practicality of the energy farm concept in that the land area required for biomass production must be very large.

Area requirements based on sea farming are small in comparison to the vast water surface area available. Assuming methane generation can be

TABLE 10.1 Area Requirements for a Marine Biomass
Farm System

Gas Supply Via Marine Biomass ($ft^3 \times 10^{12}$)	Percentage of Present Demand	Area Required (acres)	Fraction of Ocean Area
0.22	1	220×10^4	2×10^{-5}
2.20	10	220×10^5	2×10^{-4}
11.00	50	110×10^6	1×10^{-3}
22.00	100	220×10^6	2×10^{-3}

Source: From Ref. 183.

carried out at a yield of 100,000 SCF/acre per year from a marine farm, the surface area requirements to meet various percentages of the U.S. demand for natural gas as of 1977 have been computed (see Table 10.1). From this example it can be appreciated that the acquisition of necessary arable land for large-scale production of land biomass would pose a formidable problem. However, area requirements are relatively insignificant fractions of the ocean's total surface. On land, the requirement for arable soil must compete with requirements for food and fiber crops as well as the need for reserved areas for residential and recreational uses. In addition, the requirements for large amounts of fresh water for irrigation purposes make this approach less economically attractive than the marine alternative.

Mariculture, or sea farming, may be defined as any technology used to rear, transplant, breed, or culture marine animal protein in brackish or coastal waters for the specific purpose of providing food and/or energy to man. The overall coverage of this subject is summarized in Table 10.2.

Mariculture involves the areas of marine biology, marine husbandry, coastal engineering, economics, oceanography, nutrition, genetics, management, and chemistry. Marine husbandry might be considered to be the most important because it is the link between the technology of "wild" fish acquisition and the technology of domesticated marine organisms. The scientific theory of mariculture may very well rest with the marine biologist, but the application of the scientific theory by means of technology requires a combination of general skills in animal culture and marine husbandry, nutrition, coastal fluid dynamics, economics, and good management.

TABLE 10.2 Maricultural Technology and Related Considerations

I. Physical and Functional Purposes	

A. Mariculture scope
  1. Rearing marine animals
  2. Transplanting marine
     animals
  3. Breeding marine animals
  4. Culturing marine animals
	E. Products affected
	  1. Input products
	     a. Fertilizer
	     b. Feed
	     c. Boats
	     d. Pumps and chemicals

B. Scientific disciplines involved
  1. Marine biology
  2. Marine husbandry
  3. Coastal engineering
  4. Economics
  5. Oceanography
  6. Chemistry
  7. Nutrition
	2. Output products
	     a. Fresh fish
	     b. Processed fish
	     c. Consumer prepared
	        fish foods
	
	F. Design-dimension data
	   Coastal and brackish zones

C. Industries involved
  1. Fishing
  2. Agriculture
  3. Marine supply
  4. Dredging and construction
  5. Pipe making
  6. Chemical manufacturing
  7. Fertilizer
  8. Pump manufacturing
  9. Laboratory supplies
  10. Boat making
	G. Manufactured components
	  1. Pumps
	  2. Prefabricated ponds
	  3. Boats
	  4. Marine supplies
	  5. Chemicals

D. Professions and occupations
   involved
  1. Finance and banking
  2. Lawyers
  3. Boat operators
  4. Marine biologists
  5. Entrepreneurs
  6. Economists
  7. Fisherman
  8. Nutritionists	

Table 10.2 (continued)

II. Present State of the Art

A. Current state of the assessed technology
   1. Limited application in Japan, Great Britain, Russia, and most countries of Southeast Asia
   2. Present annual production 3.5 million tons
   3. Mostly luxury crop production—oysters, shrimp, mussels, etc.
   4. Sustained yields are about 0.5 ton/acre

B. Current state of supporting sciences and industry
   1. Supporting transportation needs further application
   2. Supporting agriculture is just moving from subsistence orientation to market orientation
   3. Wild fishing same for the last 4000 years
   4. Water pollution control is just in its infancy but growing rapidly
   5. Food preservation still needs a cheaper alternative than refrigeration—possibly irradiation
   6. Marine ecology is still in its infancy

III. Factors Influencing Technology Development

A. Technical breakthroughs needed for development and application
   1. Predator controls
   2. Disease and virus controls
   3. Cannibalism control
   4. Understanding of species life cycles
   5. Location and species selection criteria
   6. Cheap feed formula
   7. "Cook book" methodology
   8. Pesticide and herbicide pollution control

B. Institutional factors affecting development and application
   1. Research money
   2. Investment funds and foreign aid
   3. Training programs for technical assistance
   4. Laws pertaining to nearby waters and rights on the continental shelf as well as the open ocean
   5. Governmental action
   6. Growing nationalism and international relations

(continued)

Table 10.2 (continued)

C. Industrial vs. consumer markets
   Overwhelmingly consumer oriented

D. Buyers
   1. Luxury crops to developed countries (North America, Europe, and Australia)
   2. Nonluxury crops to developing countries (Asia, Africa, and South America)

E. Marketing channels
   Conventional food distribution industry

F. Financing
   1. Foreign aid (bilateral), 20%
   2. Foreign aid (multilateral), 40%
   3. Private investment, 40%

IV. Technologies Related to Mariculture

Supporting technologies
1. Agriculture
2. Fishing
3. Water pollution control
4. Coastal engineering
5. Food processing
6. Food distribution and storage

POTENTIAL OF SEA FARMING

A number of research programs are currently underway to investigate the potential of various marine biomass sources as fuels (Fig. 10.1) as well as the concept of large-scale mariculture. The giant brown kelp Macro-cystis pyrifera was briefly discussed earlier as one marine biomass show-ing great promise for methane generation. This species is the largest of the marine algae, attaining lengths of some 200 ft in adult plants. Figure 10.2 illustrates the basic parts of a young adult Macrocystis plant. This particular species is easily cultured and has been observed to attain growth rates as high as 20% per day under laboratory conditions (184).

The American Gas Association (AGA) is attempting to develop various substrate structure configurations to serve as support structures for growth. One substrate configuration being considered in the AGA study is illustrated in Fig. 10.3A. The system shown is a tension grid structure. A quarter acre module with this substrate configuration is designed to be moored in water as shown in Fig. 10.3B. The structure is so designed that the base of the kelp plant is fixed at a specified depth.

At present, giant kelp is harvested commercially at a rate of about 130,000 tons/year. Harvesting is done with specialized vessels specifically designed for this purpose. The kelp is usually cut at a depth of about 3 or

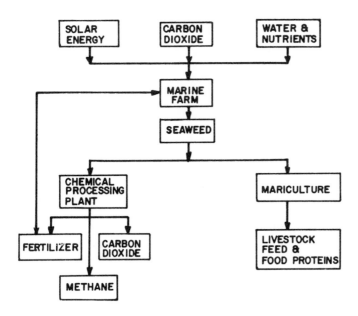

FIG. 10.1 Scheme for generating methane from marine biomass.

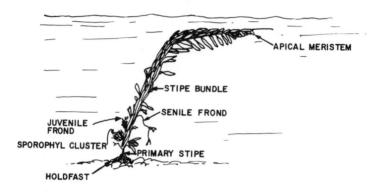

FIG. 10.2 The basic parts of a young adult giant kelp.

4 ft below the surface and unloaded into the hold of the vessel by a conveyor.
Bryce (183) notes that for the marine farm concept, harvested plants could
be further processed aboard ship by chopping and grinding them to increase
the amount of material that could be transported and to facilitate unloading
at dockside.

Raw kelp consists of roughly 87% water. This may prove to be advantageous as in the case of anaerobic digestion, as the raw material could be
introduced into the digesters with little or no addition of process water.
Table 10.3 gives a breakdown of the typical composition of giant kelp. After
anaerobic digestion and dewatering, solids are suitable for use as an animal
feed supplement and/or fertilizer. Effluent fluids are suitable for use as
fertilizer for land crops, or they can be returned to the marine farm as
fertilizer for the kelp.

PHOTOSYNTHETIC HYDROGEN PRODUCTION
SYSTEMS AND COMPETING MARKETS

Algae and photosynthetic bacteria have been recognized for their potential
large-scale hydrogen-producing capability since the 1940s. Figure 10.4
illustrates the mechanism of hydrogen evolution from marine microorganisms. Solar radiation is absorbed by some form of pigment. Energy is
transmitted to an energized electron through the reaction-center chlorophyll
protein. Free electrons result from hydrogen donated by some electron
donor such as water. The energized electrons for hydrogen production are
visualized as following the same pathway as those leading to carbon fixation.
The electrons are shunted from the photosynthetic-carbon fixation pathway
to ferredoxin and are then transferred from ferredoxin to H^+. This last
step in hydrogen gas production is usually catalyzed by an enzyme (i.e.,

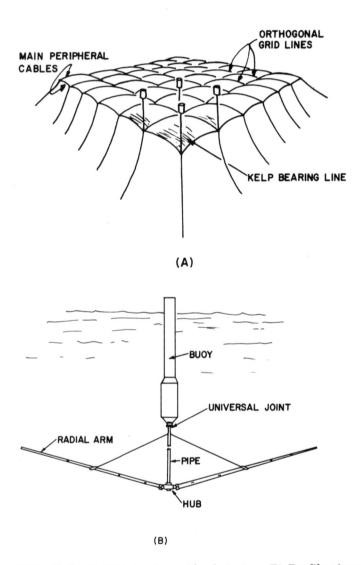

(A)

(B)

FIG. 10.3 (A) AGA tension grid substrate.   (B) Profile view of the quarter acre module.   (From Ref. 183.)

TABLE 10.3 Typical Composition of Macrocystis

Ash	38-44
KCl	28.7
NaCl	7.5
Na_2SO_4	4.3
Trace	4.0
Volatile solids	56-62
Carbohydrates	26.0
Mannitol	0-14
Laminarin	0-21
Fucoidan	0.5-2
Structure and pigments	27.5
Algin	13-24
Cellulose	3-8
Protein	5-7
Fat	0.5

hydrogenase or nitrogenase). Mitsui (185) describes the mechanism in detail.

There are two approaches to improving the economic feasibility of a hydrogen-producing system: one involves maximizing the solar conversion efficiency; the other requires development of profitable ways to use the by-products of the reaction, thereby enhancing the net worth of the system.

Because the production of hydrogen gas is a by-product of the photo-synthetic process, solar conversion efficiency is directly related to the metabolic requirements of the plant cells (186). This means that the prob-lem of increasing bioconversion efficiency involves a complex of parameters that are associated with photosynthesis as well as plant metabolism. Mitsui (185) notes that the following factors further complicate the problem of increasing photosynthetic efficiency:

1. The growth rates of plants are directly related to the species' photo-synthetic capability. This in turn determines the hydrogen-producing potential of the biomass. Different plant species have different growth potentials; even within the same species, attainment of maxi-

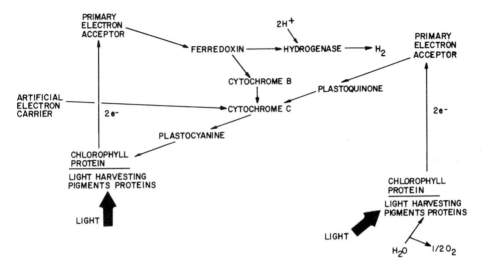

FIG. 10.4   The mechanism behind hydrogen production in algae.   (From Ref. 185.)

mum production is contingent on proper environmental conditions (i.e., sunlight, temperature, salinity, nutrient availability, etc.).

2. Electrons may be shared among competing pathways. That is, there are numerous pathways which the excited electron can take, including CO_2 fixation, O_2 reduction, cyclic electron flow, nitrogen fixation, nitrite reduction, or hydrogen production. The metabolic state of the plant cell will determine the portion of electrons flowing along each of these pathways.

3. Another consideration is that in the natural environment, oxygen generated during the photolysis of water can inhibit the activities of some types of hydrogenase enzymes which otherwise would catalyze the H_2-producing and N_2-fixing reactions. This will reduce the net yield of hydrogen gas.

4. Certain steps in the transfer of electrons to hydrogen generation are rate limiting.

There are a number of ways in which hydrogen production systems could be employed in multiple utilization schemes. Most of these were discussed earlier. By way of review, the following large-scale applications show potential:

1. Various species of blue-green and green algae as well as some photosynthetic bacteria are highly rich in protein. After mass cul-

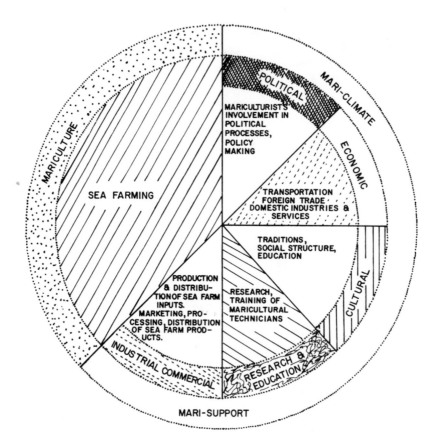

FIG. 10.5 The major factors affecting sea farming. (From Ref. 187.)

tures have been used for hydrogen production, the harvested protein
might serve as a source of food for human and animal consumption.
Still another possibility is to use it as a food source for aquacultural
farms (i.e., fish, shellfish, shrimp, etc.). Algal carbohydrates
have the potential of being converted to alcohol and methane gas.
2. Mass cultures of algae and photosynthetic bacteria could also be
    utilized as sources of organic substrates or methane-producing sub-
    stances.
3. Algae and photosynthetic bacteria are also potential sources of new
    pharmaceutical chemicals.

    A broad definition of mariculture should include the support activities
that will provide sea farms with production inputs and those industries that

will market, process, and distribute maricultural products. Thus, the
three main components of marine farming can be considered to be mariculture, "mari-support," and "mari-climate." Figure 10.5 illustrates
these areas and their interrelationship.

THE NATURE OF SEA FARMING

It is of interest to contrast mariculture with agriculture. Both agriculture
and mariculture require an extensive and well-articulated transportation
system to move the production inputs from distant points of manufacture to
each farm—on land or sea—and to move products to the ultimate consumer.

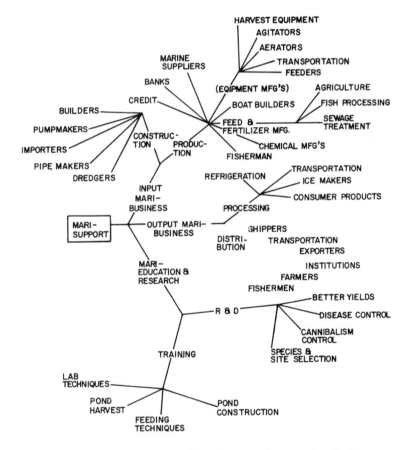

FIG. 10.6 Specific factors influencing mari-support activities.

But mariculture, unlike agriculture, can avoid seasonal variations and
crop-rotation schemes, and can thus avoid intermittent labor-supply
problems.

Mariculture and agriculture are similar (and like many other technologies)
in that their productive capacity is improvable by human innovation.

Agriculture can be efficient at the small-farm level; however, mariculture tends to be more efficient as size increases up to some optimum size
(187). Both land farms and sea farms can be too large to be efficient if
the size of the farm removes the incentive to increase productivity per acre,
or where complexity of the management problem exceeds the technological
means for meeting it.

One of the basic changes that will characterize developing mariculture
as it grows in productivity is that the sea farm will come to resemble the
assembly line in factory manufacturing enterprises (i.e., intensive culture).
Such industrial-type sea farms might be expected to depend less and less
on resources inherent in the water and more and more on production inputs
purchased from other sectors of the economy. It is likely that such farming
will be serviced more and more by diverse mari-support activities.

Sea farming is thus likely to become interrelated with many other enterprises and technologies. These will service both industrial and commercial
needs of the maricultural industry. Government may also provide some
initial financing and other aid (e.g., research funding). Many of these
operations will be carried on by large support organizations that serve sea
farm businesses but are not part of them, thereby permitting efficient
economies of scale. Figure 10.6 illustrates some of the many elements
involved in this mari-support.

The production of fertilizers and sea farm equipment (pumps, pipes,
dredges, boats, etc.) is normally economical only when conducted on a
large scale, i.e., in plants requiring heavy capital investment. The manufacture of fertilizers, for example, requires large amounts of power and
entails heavy expenditures for transportation of raw materials. Formulating
maricultural feeds will become increasingly important as mariculture pro-
gresses. This feed industry will utilize by-products of other industrial
processes.

Research will be essential to the successful economic development of
marine biomass. For sea farming to become more efficient and increase
its output, major advances in technology will be necessary. Only in a
favorable politico-economic climate will research be supported that can
produce steady development in technology yielding improved marine feedstocks and equipment as well as better bioconversion techniques.

Glossary

Aerobic. In the presence of air or oxygen.

Anaerobic. In the absence of air or oxygen.

Biogas. Gaseous product generated by anaerobic fermentation of organics; Methane is chief constituent.

Boiler. Closed pressure vessel in which a liquid is vaporized by the application of heat.

British thermal unit (Btu). Mean British thermal unit is 1/180 of the heat needed to raise the temperature of 1 lb of water from 32° F to 212° F at constant atmospheric pressure.

Capacity factor. Ratio of the average load on a machine for the period of time considered to the capacity rating of the machine.

Capital costs. Investment costs needed to construct a system.

Cellulose. Biological polymer consisting of sugar molecules; basic material of plant fiber.

Cellulolytic. Having the property of hydrolyzing cellulose.

Coal gas. Gas formed from destructive distillation of coal.

Coal tar. Black, highly viscous liquor that is a by-product formed from the distillation of coal.

Coke. Fuel composed primarily of fixed carbon and ash in coal, generated from the destructive distillation of bituminous coal.

Coke oven gas. Gas generated via destructive distillation of bituminous coal in closed chambers.

197

Coking. Conversion of carbonaceous fuel by heating in absence of air.

Combustion. Incineration; the chemical combination of oxygen with combustible constituents of a fuel, resulting in heat.

Conventional fuels. Fossil fuels (e.g., coal, gas, oil).

Cracking. Thermal decomposition of hydrocarbons into simpler compounds.

Crude oil. Unrefined petroleum.

DOE. U.S. Department of Energy, a federal cabinet-level department created in 1977.

Detention time. Time that a material remains in a reactor or system; defined as the total weight of material in the system divided by the total weight removed per unit time (units of time); in this volume this term is used interchangeably with retention time.

Digestion. Decomposition of complex organic molecules into simpler molecular structures; in an anaerobic process, referred to as fermentation, in which microorganisms accomplish the decomposition.

Dry fuel basis. Method of reporting fuel analysis without moisture or other constituents.

DTE. Dry ton equivalents.

Enzyme. Biological catalyst that speeds up the breakdown of complex organic molecules into simpler structures.

ERDA. Energy Research and Development Administration; now incorporated into the U.S. Department of Energy (DOE).

Fermentation. Biological process in which organics are decomposed into simpler constituents by microorganisms.

Generating station. Plant consisting of prime movers, electric generators, and auxiliary equipment for converting mechanical, chemical, or nuclear energy into electric energy.

Generator, steam. Machinery that burns fuel and transforms water to steam.

Heat exchangers. Device that transfers heat from one fluid to another.

kW. Kilowatt (10^3 watts), a measure of power.

kWhr. Kilowatt-hour, a measure of energy.

Lipids. Fatty materials.

Lipolytic. Having the property of hydrolyzing lipid materials.

Liquefied petroleum gas (LPG). Consists of hydrocarbons that are normally gases at atmospheric conditions but can be liquefied under moderate pressures; derived from natural gas and refinery sources via crude distillation and cracking.

Methanogenic. Having the property of producing methane.

MW. Megawatt (10^6 watts); a measure of power.

Natural gas. Naturally occurring gaseous fuel consisting largely of methane.

NASA. National Aeronautics and Space Administration.

Night soil. Human feces.

Petroleum. Naturally occurring mineral oil largely composed of hydrocarbons.

Phytotoxic. Poisonous to plants.

Power. Rate at which energy is consumed [units may be kW, MW, horsepower (hp) etc.].

Proteolytic. Having the property of hydrolyzing proteins.

Refinery gas. Commercial noncondensible gas derived from fractional distillation of crude oil or cracking of crude oil on petroleum distillates.

Retention time. See Detention time.

Steam. Vapor phase of water significantly unmixed with other gases.

Substrate. Material supplied for microbial action.

Superheated steam. Steam at a higher temperature than its saturation temperature.

Transmission. Method of transferring electric energy in bulk from a source in the generating system to other utility systems.

Transmission line. Line used for bulk transmission of electric energy between a generating or receiving point and delivering stations.

Total solids. Weight of solids remaining after material is dried at constant weight at $103 \pm 1°C$.

Volatile acids. Low-molecular-weight fatty acids.

Volatile solids. Portion of solids volatilized at $550 \pm 50°C$; difference between total solids content and ash remaining after ignition at $550 \pm 50°C$.

WEC. Wind energy conversion.

WECS. Wind energy conversion system.

References

1. E. M. Wilson and H. M. Freeman, "Processing Energy from Wastes," Environ. Sci. Technol. Vol. 10 (May 1976).
2. P. N. Cheremisinoff and A. Morresi, Geothermal Energy: Technology Assessment, Technomic Publ., Westport, Conn., 1976.
3. N. P. Cheremisinoff, Fundamentals of Wind Energy, Ann Arbor Science Publ., Ann Arbor, Mich., 1978.
4. G. E. Hutchinson, "The Biosphere," Sci. Amer., Vol. 223, No. 3 (Sept. 1970).
5. G. M. Woodwell, "The Energy Cycle of the Biosphere," Scientific American, Vol. 223, No. 3 (Sept. 1970).
6. B. Bolin, "The Carbon Cycle," Sci. Amer., Vol. 223, No. 3 (Sept. 1970).
7. The Mitre Corp., Metrek Div., "Fuels from Biomass-Near-Term Initiative," WP12282 (April 7, 1977).
8. P. N. Cheremisinoff, P. P. Cheremisinoff, A. C. Morresi, and R. A. Young, Woodwastes Utilization and Disposal, Technomic Publ., Westport, Conn., 1976.
9. T. B. Reed, Free Energy of Formation of Binary Compounds, MIT Press, Cambridge, Mass., 1971.
10. T. B. Reed, "Biomass Energy Refineries for Production of Fuel and Fertilizer," Proc. 8th Cellulose Conf. (TAPPI), May 20-22, 1976, Syracuse, N.Y.
11. E. S. Lipinsky, "Fuels from Biomass: Integration with Food and Materials Systems," Science, Vol. 199 (Feb. 10, 1978).
12. A. L. Lehninger, Biochemistry, 2nd ed., Worth, New York, 1975.
13. "The Outlook for Timber in the United States," U.S. Dept. of Agriculture, Forest Service, Forest Resource Rept. No. 20 (1973).

202

References

14. P. N. Cheremisinoff and A. C. Morresi, "Energy from Wood Wastes," Environment, Vol. 19, No. 4 (1977).
15. P. N. Cheremisinoff and A. C. Morresi, Energy from Solid Wastes, Dekker, New York, 1976.
16. J. I. Zerbe and R. A. Arola, "Direct Combustion of Silvicultural Biomass," paper presented at Fuels from Biomass Symposium, Univ. of Illinois, Urbana, Ill. (April 19, 1977).
17. "Chemicals from Wood Waste," U.S. Dept. of Agriculture, Forest Service, Springfield, Va., Rept. No. AN441 (Dec. 24, 1975).
18. "Tomorrow's Feedstocks: From Where Will They Come?" Chem. Eng. Progr. (Dec. 1977).
19. G. E. Weismantel, "Can Natural Oils Feed the Plastics Industry?" Chem. Eng., Vol. 82, No. 5 (March 3, 1975).
20. J. Jones, "Converting Solid Wastes and Residues to Fuel," Chem. Eng., Vol. 85, No. 1 (Jan. 2, 1978).
21. L. R. Brown, "Human Food Production as a Process in the Biosphere," Sci. Amer., Vol. 223, No. 3 (Sept. 1970).
22. A. Mitsui, "The Use of Photosynthetic Marine Organisms in Food and Feed Production," paper presented at Int. Conf. on Bio-Saline Research, South Carolina (Sept. 1977).
23. A. Muira, "Porphyra Cultivation in Japan," in Advance of Phycology in Japan (J. Todida and H. Hirose, eds.), Funk, The Hague, 1975, pp. 273-304.
24. "Culture of Algae and Seaweeds," Food and Agriculture Organization, Rome, Fish. Rept. No. 188, pp. 34-35 (1976).
25. G. T. Velasquez, "Studies and Utilization of the Philippine Marine Algae," in Proc. 17th Int. Seaweed Symposium (K. Nisijawa et al., eds.), Wiley, New York, 1971, pp. 62-65.
26. J. Naylor, "Production, Trade and Utilization of Seaweeds and Seaweed Products," Food and Agriculture Organization, Fish. Technol. Paper No. 159 (1976).
27. Food and Agriculture Organization, Yearbook of Fisheries Statistics, Vol. 38, p. 40 (1975).
28. T. Arasaki and N. Mino, "Alkali-Soluble Proteins in Marine Algae," Eiyo To Shokuryo, Vol. 26, No. 2, pp. 129-133 (1973).
29. B. Volesky, J. E. Zajic, and E. Knettig, "Algal Properties," in Properties of Products of Algae (J. E. Azjic, ed.), Plenum Press, New York, 1970.
30. H. Gest and M. D. Kamen, "Photoproduction of Molecular Hydrogen by Rhodospirillum Rubrum," Science, Vol. 108, No. 558 (1949).
31. S. Lien and A. San Pietro, "An Enquiry into Biophotolysis of Water to Produce Hydrogen," Indiana Univ. and NSF (1975).
32. A. Mitsui and S. Kumazawa, "Hydrogen Production by Marine Photosynthetic Organisms as a Potential Energy Source," in Biological Solar Energy Conversion (A. Mitsui et al., eds.), Academic Press, New York, 1977.

33. A. Flowers and A. J. Bryce, "Energy from Marine Biomass," Sea Technol. (Oct. 1977).

34. H. S. Gordon, "A Rising Tide for Seaweed," Chem. Eng., Vol. 83, No. 26 (Dec. 6, 1976).

35. "Making Aquatic Weeds Useful: Some Perspectives for Developing Countries," National Academy of Sciences, Washington, D.C. (1976).

36. B. C. Wolverton, R. C. McDonald, and J. Gordon, "Bio-conversion of Water Hyacinths into Methane Gas: Part 1," NASA Tech. Memo. No. Tm-X-72725 (1975).

37. S. N. Singh and L. P. Sinha, "Studies on Use of Water Hyacinth Culture in Oxidation Ponds Treating Digested Sugar Wastes and Effluent of Septic Tank," Environ. Health, Vol. 11 (1969).

38. M. K. Hubbert, "The Energy Resources of the Earth," Sci. Amer., Vol. 224, No. 3 (Sept. 1971).

39. W. B. Kemp, "The Flow of Energy in a Hunting Society," Sci. Amer., Vol. 244, No. 3 (Sept. 1971).

40. E. Cook, "The Flow of Energy in an Industrial Society," Sci. Amer., Vol. 224, No. 3 (Sept. 1971).

41. Statistical Yearbook, 1971, United Nations Statistical Office, N.Y. (1972).

42. H. A. Wilcox, "The Energy Growth: Present Trends and Future Prospects for the World and the USA," paper presented to the Marine Technology Soc., Washington, D.C. (June 12, 1973).

43. H. Kahn, "Prospects for Mankind," 1973 Synoptic Context on the Corporate Environment: 1975-1985, Vol. 2, Hudson Inst., Croton-on-Hudson, N.Y., 1973.

44. D. Chapmann et al., Science, Vol. 178, No. 4062 (Nov. 17, 1972).

45. C. A. Zraket, "Growth and the Conservation of Energy," MTP-389, The Mitre Corp (Feb. 1974).

46. R. S. Greeley, "Energy Use and Climate," The Mitre Corp., M74-66 (April 1975).

47. National Petroleum Council, "U.S. Energy Outlook: An Initial Appraisal 1971-1985" (July 1, 1971).

48. M. R. Gustavson, "Dimensions of World Energy," The Mitre Corp., M71-71 (Nov. 1971).

49. E. Teller, paper presented at Symposium on Energy, Resources and the Environment, The Mitre Corp., M72-154 (May 1972).

50. P. N. Cheremisinoff and T. C. Regino, Principles and Applications of Solar Energy, Ann Arbor Sci. Publ., Ann Arbor, Mich., 1978.

51. "Energy Alternatives: A Comprehensive Analysis," report prepared for Council on Environmental Quality/NSF-Office of Energy R&D Policy, by Univ. of Oklahoma, Norman, Okla. (1975).

52. G. M. Woodwell, "The Carbon Dioxide Question," Sci. Amer., Vol. 238, No. 1 (Jan. 1978).

53. W. W. Kellogg, "Climate Change and the Influence on Man's Activities on the Global Environment," M72-166, The Mitre Corp. (Sept. 1972).
54. H. Flohn, Climate and Weather, McGraw-Hill, New York, 1969.
55. E. D. Sellers, Physical Climatology, Univ. of Chicago Press, Chicago, 1965.
56. M. I. Budyko, "Comments on a Global Climate Model Based on the Energy Balance of the Earth-Atmosphere System," J. Appl. Meteorol. Vol. 9 (1970).
57. W. W. Kellogg, "Long Range Influences of Mankind on the Climate," paper presented at the World Conf. "Toward a Plan of Action for Mankind," Inst. de la Vie, Paris, France (Sept. 9, 1974).
58. C. E. P. Brooks, Climate Through the Ages, 2nd ed., Dover, New York, 1970.
59. "Living with Climatic Change, Phase II," Symposium Proc., The Mitre Corp./Metrek Div., MTR-7443 (June 1977).
60. "An Assessment of Solar Energy as a National Energy Resource," NSF/NASA Solar Energy Panel, Univ. of Maryland, College Park, Md. (Dec. 1972).
61. K. Howlett and A. Gamache, "Forest and Mill Residues as Potential Sources of Biomass," Silvicultural Biomass Farms, Vol. 6, The Mitre Corp., Metrek Div., Tech. Report No. 7347 (May 1977).
62. R. C. Hawley and D. C. Smith, The Practice of Silviculture, Wiley, New York, 1954.
63. H. E. Young, "The Response of Loblolly and Slash Pines to Phosphate Manures," Queensland J. Agr. Sci., Vol. 5 (1948).
64. H. E. Young, "Complete-Tree Concept: 1964-1974," Forest Products J., Vol. 24, No. 12 (1974).
65. F. B. Golley, "Energy Values of Ecological Materials," Ecology, Vol. 42 (July 1961).
66. G. C. Szego and C. C. Kemp, "Energy Forests and Fuel Plantations," Chemtech, Vol. 3, No. 5 (1973).
67. H. Resch, "The Physical Energy Potential of Wood," in Loggers Handbook, Vol. 35, Sect. 2, pp. 29-33 (1975).
68. J. M. Harkin and J. W. Rowe, "Bark and Its Possible Uses," U.S. Dept. of Agriculture, Forest Service, Research Note FPL-091, Forest Products Lab., Madison, Wisc. (1971).
69. P. Koch, "Utilization of the Southern Pines," U.S. Dept. of Agriculture, Forest Service, Agricultural Handbook No. 420, Southern For. Exp. Sta. (1972).
70. A. J. Panshin et al., Forest Products, McGraw-Hill, New York, 1962.
71. "The Outlook for Timber in the United States," U.S. Dept. of Agriculture, Forest Service, Forest Resource Rept. No. 20, Washington, D.C. (1973).
72. J. L. Yount and R. A. Crossman, Jr., "Eutrophication Control by Plant Harvesting," J. Water Poll. Control Fed., Vol. 42 (1970).

73. M. G. McGarry and C. Tongkasame, "Water Reclamation and Algae Harvesting," Water Poll. Control Fed., Vol. 43 (1971).

74. M. McGarry, "Water and Protein Reclamation from Sewage," Process Biochem., Vol. 6 (1971).

75. E. C. Knapp, "Agriculture Poses Waste Problems," Environ. Sci. Technol., Vol. 4 (1970).

76. "Cattle, Sheep, and Goat Inventory," Statistical Reporting Service of U.S. Dept. of Agriculture, Pub. No. LV-GN-1(72) (Jan. 1972).

77. C. G. Bolueke and P. H. McGauhey, "Comprehensive Studies of Solid Waste Managements First Annual Report," Research Grant EC-00260, Univ. of California, U.S. Dept. of Health, Education and Welfare, Bureau of Solid Waste Management (1970).

78. Division of Solar Energy, "Strategy for the Fuels from Biomass Program," ERDA, Washington, D.C. (June 27, 1977).

79. G. C. Szego and C. C. Kemp, "Energy Forests and Fuel Plantations," Chemtech (May 1973).

80. G. C. Szego and C. C. Kemp, "The Estimated Availability of Resources for Large-Scale Production of SNG by Anaerobic Digestion of Specially Grown Plant Material," InterTechnology, Project No. 011075, Warrenton, Va. (1975).

81. J. A. Alich, Jr., and R. E. Inman, "Effective Utilization of Solar Energy to Produce Clean Fuel," Stanford Research Inst. (NSF/RANN/ SE/GI/38723), SRI Project No. 2643, Menlo Park, Calif. (1974).

82. R. E. Inman, D. J. Salo, and B. J. McGurk, "Silvicultural Biomass Farms: Site Specific Production Studies and Cost Analyses," Vol. 4, The Mitre Corp., Tech. Rept. No. 7347 (May 1977).

83. D. Blake and D. Salo, "Systems Descriptions and Engineering Costs for Solar-Related Technologies-Biomass Fuels Production and Conversion Systems," Vol. 9, The Mitre Corp., Tech. Rept. No. MTR-7485 (June 1977).

84. J. Warner, Hawaiian Agronomics Co., Fuels from Sugar Crops Tutorial Conf., Columbus, Ohio (Oct. 13-15, 1976).

85. E. S. Lipinsky, R. A. Nathan, W. J. Sheppard, T. A. McClure, W. T. Lawhon, and J. L. Otis, "Systems Study of Fuels from Sugarcane, Sweet Sorghum, and Sugar Beets: Comprehensive Evaluation," Battelle Columbus Labs., Columbus, Ohio, BMI-1957: Vol. 1 (1977).

86. R. Dideriksen, A. Hidlebaugh, and K. Schmude, "Potential Cropland Study," SCS, U.S. Dept. of Agriculture, Washington, D.C. (1977).

87. Agricultural Statistics: 1976, U.S. Department of Agriculture, available from U.S. Gov. Printing Office, Washington, D.C. (1976).

88. An Evaluation of the Use of Agricultural Residues as an Energy Feedstock, Vol. 1, Stanford Research Inst., Menlo Park, Calif., 1976

89. "Biomass: A Cash Crop for the Future?" Conf. on the Production of Biomass from Grains, Crop Residues, Forages and Grasses for Conversion to Fuels and Chemicals, Kansas City, Mo. (Mar. 1977).

90. "A Feasibility Study on the Use of Crop Residue, Feedlot Waste, and Municipal Waste to Support a Municipal Electric Utility," Progr. Rept. Eng. Exp. Station, Kansas State Univ., Manhattan, Kans. (Oct. 1976).

91. Crop, Forestry and Manure Residue Inventory—Continental United States, Stanford Research Inst., Menlo Park, Calif. (1976).

92. M. L. Cotner, M. D. Skold, and O. Krause, "Farmland: Will There be Enough?" U.S. Dept. of Agriculture, Rept. No. ER5-584, Washington, D.C. (1975).

93. T. W. Jeffries, P. H. Moulthrop, H. Timourian, R. L. Ward, and B. J. Berger, Biosolar Production of Fuels from Algae, Lawrence Livermore Labs., Livermore, Calif., UCRL-52177 (1976).

94. V. S. Budhraja, B. N. Anderson, G. A. Hoffman, F. J. Nickels, R. H. Schneider, and D. H. Walsh, "Ocean Food and Energy Farm Project. Subtask No. 6: Systems Analysis," ERDA/USN/1027-76/1 (Vol. 1), Integrated Sciences Corp., Santa Monica, Calif. (1976).

95. W. J. Oswald and C. G. Golueke, "Biological Transformation of Solar Energy," Advan. Appl. Microbiol., Vol. 2, No. 223 (1960).

96. G. E. Fogg, "Physiology and Ecology of Marine Blue-Green Algae," in Biology of Blue-Green Algae (N. G. Carr and B. A. Whitton, eds.), Botanical Monographs, Vol. 9, Univ. of California Press, Berkeley, Calif., 1973.

97. R. W. Castenholz, "Thermophilic Blue-Green Algae and the Thermal Environment," Bacteriol. Rev., Vol. 33, No. 476 (1969).

98. L. L. Anderson, "Energy Potential from Organic Wastes: A Review of the Quantities and Sources," U.S. Dept. of the Interior, Bur. of Mines Information Circ. No. 8549 (1972).

99. D. J. DeRenzo, Energy from Bioconversion of Waste Materials, Noyes Data Corp., Park Ridge, N.J., 1977.

100. J. R. Alich, Jr., and J. G. Witwer, "Agricultural and Forestry Wastes as an Energy Resource," Solar Energy, Vol. 19 (1977).

101. R. S. Greeley and P. C. Spewak, "Land and Fresh Water Energy Farming," The Mitre Corp., presented at Washington Center for Metropolitan Studies Bioconversion Conf. (Mar. 10, 1976).

102. J. Alich and R. Inman, "Energy from Agriculture," paper presented at Conf. on Clean Fuels from Biomass, Sewage, Urban Refuse, and Agricultural Wastes, Orlando, Fla. (Jan. 26-30, 1976).

103. "California Project Would Use Agricultural Waste as Power," Christian Science Monitor, Vol. 12, No. 28 (Dec. 30, 1977).

104. T. R. Schneider, "Substitute Natural Gas from Organic Materials," paper presented at A.S.M.I. Winter Annual Meeting, New York (Nov. 1972).

105. G. Bykinsky, "A New Scientific Effort to Boost Food Output," Fortune, Vol. 91, No. 6 (June 1975).

106. E. H. White and D. D. Hook, "Establishment and Regeneration of Silage Plantings," Iowa State J. Res., Vol. 49, No. 3 (1975).

107. H. J. Kitzmiller, Jr., "Response of Sycamore Families to Nitrogen Fertilization," Tech. Rept. No. 48, School of Forest Resources, North Carolina State Univ., Raleigh, N.C. (1972).

108. F. T. Bonner and W. M. Broadfoot, "Growth Response of Eastern Cottonwood of Nutrients in Sand Culture," U.S. Dept. of Agriculture, Forest Res. Note SO-65, Southern For. Exp. Station, New Orleans (1967).

109. P. E. Heilman, D. V. Peabody, Jr., D. S. DeBell, and R. F. Strand, "A Test of Close-Spaced, Short-Rotation Culture of Black Cottonwood," Can. J. For. Res., Vol. 2, No. 4 (1972).

110. R. Hunt, presentation at Eucalyptus Workshop, Bainbridge, Ga. (Sept. 1, 2, 1976).

111. R. Hunt and B. Zobel, "Early Growth and Future Possibilities of Eucalyptus in the Southeast Coastal Plain," Alabama For. Products, Vol. 17, No. 8 (1974).

112. W. L. Pritchett and W. H. Smith, "Fertilizing Slash Pine on Sandy Soils of the Lower Coastal Plain," in Tree Growth and Forest Soils (C. T. Youngberg and C. T. Davey, eds.), Oregon State Univ. Press, Corvallis, Ore., 1970.

113. W. L. Pritchett and W. H. Smith, "Management of Wet Savannah Forest Soils for Pine Production," Florida Agr. Exp. Sta. Tech. Bull. (1973).

114. R. P. Schutz, "Intensive Culture of Southern Pines: Maximum Yields on Short-Rotations," Iowa State J. Res., Vol. 49, No. 3 (1975).

115. R. B. Heiligmann, "Weed Control for the Intensive Culture of Short-Rotation Forest Crops," Iowa State J. Res., Vol. 49, No. 3 (1975).

116. W. Oswald, "Gas Production from Micro Algae," paper presented at Conf. on Clean Fuels from Biomass, Sewage, Urban Refuse, and Agricultural Wastes, Orlando, Fla. (Jan. 26-30, 1976).

117. M. McGarry, "Water and Protein Reclamation from Sewage," Process Biochem., Vol. 6, No. 1 (Jan. 1971).

118. J. Yount and R. Crossman, "Eutrophication Control by Plant Harvesting," J. Water Poll. Cont. Fed. (May 1970).

119. A. Mitsui, "Long Range Concepts: Applications of Photosynthetic Hydrogen Production and Nitrogen Fixation Research," paper presented at Capturing the Sun through Bioconversion Conf., Washington, D.C. (Mar. 10-12, 1976).

120. R. Lecuyer, "An Economic Assessment of Fuelgas from Water Hyacinths," paper presented at Conf. on Clean Fuels from Biomass, Sewage, Urban Refuse, and Agricultural Wastes, Orlando, Fla. (Jan. 26-30, 1976).

121. "Methane Generation from Human, Animal, and Agricultural Wastes," National Academy of Sciences, Washington, D.C. (1977).

122. C. E. Fogg, "Livestock Waste Management and the Conservation Plan," in Livestock Waste Management and Pollution Abatement:

The Proceedings of the International Symposium on Livestock Wastes,
Ohio State Univ., Columbus, Ohio (April 19-22, 1971).

123. J. I. Zerbe and R. A. Arola, "Direct Combustion of Silvicultural
      Biomass," paper presented at Fuels from Biomass Symposium,
      Univ. of Illinois, Urbana, Ill. (April 19, 1977).

124. C. Bliss and D. O. Blake, "Silvicultural Biomass Farms: Conversion
      Processes and Costs," MITRE Tech. Rept. No. 7347, Vol. 5, The
      Mitre Corp., Metrek Div. (May 1977).

125. G. R. Fryling (ed.), Combustion Engineering, Combustion Engineer-
      ing, Inc., New York, 1966.

126. E. H. Hall, C. M. Allen, D. A. Ball, J. E. Burch, H. N. Conkle,
      W. T. Lawhon, T. J. Thomas, and G. R. Smithson, Jr., "Compari-
      son of Fossil and Wood Fuels," EPA Rept. No. EPA-600/2-76-056
      (March 1976).

127. K. P. Ananth, L. J. Shannon, and M. P. Schrag, "Environmental
      Assessment of Waste-to-Energy Processes, Source Assessment
      Document," EPA Rept. No. EPA-600/7-77-091 (Aug. 1977).

128. R. E. Inman, J. G. Leigh, and D. J. Serlo, "Fuels from Biomass
      Near-Term Initiative," Rept. WP-12282, The Mitre Corp., Metrek
      Div., Washington, D.C. (April 7, 1977).

129. P. G. Gorman, L. J. Shannon, M. P. Schrag, and D. Fiscus,
      "St. Louis Demonstration Project Final Report: Power Plant Equip-
      ment Facilities and Environmental Evaluation," Vol. 2, EPA Contract
      No. 68-02-1871 (July 1976).

130. "First Report to Congress: Resource Recovery and Source Reduction,"
      U.S. Environmental Protection Agency, Office of Solid Waste Manage-
      ment Programs (Feb. 22, 1973).

131. "Problems and Opportunities in the Management of Combustible Solid
      Waste," International Research & Technology Corp., Final Report
      on Contract No. 68-03-0060 to U.S. Environmental Protection Agency,
      Solid and Hazardous Waste Research Lab., NERC, Cincinnati (1972).

132. R. A. Chapman, "Solid Waste as a Fuel for Power Generation," in
      Proc. 1973 Washington State Univ. Thermal Power Conference
      (Oct. 3-5, 1973).

133. "Present Status of Methane Gas Utilization as a Rural Fuel in Korea,"
      Inst. of Agricultural Engineering and Utilization, Office of Rural
      Development, Suwon, Korea (1973).

134. L. R. Maki, "Experiments on the Microbiology of Cellulose Decom-
      position in a Municipal Sewage Treatment Plant," Antonie van Leeuwen-
      hock J. Microbiol. Serol., Vol. 20 (1954).

135. H. Heukelekian, "Decomposition of Cellulose in Fresh Sewage Solids,"
      Ind. Eng. Chem., Vol. 19 (1927).

136. R. H. McBee, "The Culture and Physiology of a Thermophilic Cellu-
      lose Fermenting Bacterium," J. Bacteriology, Vol. 56 (1948).

137. D. W. Stranks, "Microbiological Utilization of Cellulose and Wood:
     I. Laboratory Fermentations of Cellulose by Rumen Organisms,"
     Can. J. Microbiol., Vol. 2 (1956).
138. "Methane Generation from Human, Animal, and Agricultural Wastes,
     National Academy of Sciences, Washington, D.C. (1977).
139. C. Bisselle, M. Kornreich, M. Scholl, and P. Spewak, "Urban Trash
     Methanation Background for a Proof-of-Concept Experiment," MITRE
     Tech. Rept. MTR-6856, NSF-RA-N-75-002, The Mitre Corp. (Feb.
     1975).
140. "Technology for the Conversion of Solar Energy to Fuel Gas," National
     Center for Energy Management and Power, Univ. of Pennsylvania,
     Philadelphia (Jan. 1973).
141. P. L. McCarty, "Anaerobic Waste Treatment: Fundamentals, Part 2:
     Environmental Requirements and Control," Publ. Works, Vol. 95
     (Oct. 1964).
142. J. T. Pfeffer and J. C. Liebman, "Biological Conversion of Organic
     Refuse to Methane," Rept. No. NSF/RANN/SE/G1-39191/PR74/2
     (July 1974).
143. M. Wolf and R. Keenan, "Conversion of Solar Energy to Fuel Gas,"
     paper presented at Bioconversion Energy Research Conf. at Univ. of
     Massachusetts, Amherst, Mass. (June 25-26, 1973).
144. L. E. Casida, Industrial Microbiology, Wiley, New York, 1968.
145. P. L. McCarty, "Anaerobic Waste Treatment Fundamentals, Part 3:
     Toxic Materials and Their Control," Publ. Works, Vol. 95 (Nov.
     1964).
146. W. W. Eckenfelder, "Mechanisms of Sludge Digestion," Water and
     Sewage Works, Vol. 114, No. 6 (June 1967).
147. J. T. Pfeffer, "Reclamation of Energy from Organic Refuse,"
     NSF/RANN (April 1973).
148. S. A. Waksman, Soil Microbiology, Wiley, New York, 1952.
149. R. A. Boettcher, "Air Classification of Solid Wastes," U.S. Environ-
     mental Protection Agency, Office of Solid Waste Management (1972).
150. G. L. M. Christopher, "Biological Production of Methane from
     Organic Materials," Final Rept. to Columbia Gas Service System
     Corp. (1971).
151. C. N. Acharya, "Your Home Needs a Gas Plant," Indian Farming,
     Vol. 6, No. 2 (1956).
152. R. B. Singh, "Bio-Gas Plant: Generating Methane From Organic
     Wastes," Gobar Gas Research Sta., Ajitmal, Etawah (U.P.), India
     (1971).
153. T. B. Reed and R. M. Lerner, "Methanol: A Versatile Fuel for
     Immediate Use," Science, Vol. 182, p. 4119 (Dec. 28, 1973).
154. D. L. Hagen, "Methanol: Its Synthesis, Use as a Fuel, Economics,
     and Hazards," M.S. thesis, Univ. of Minnesota (Dec. 1976).

155. A. E. Hokanson and R. M. Rowell, "Methanol from Wood Waste:
     A Technical and Economic Study," U.S. Dept. of Agriculture, Forest
     Service, Gen. Tech. Rept. No. FPL-12 (June 1977).
156. J. E. Anderson, Solid Refuse Disposal Process and Apparatus,
     U.S. Patent No. 3,729,298 (April 24, 1973).
157. T. B. Reed, "Synthetic Alcohol for Fuel," statement before the
     Proxmire Committee on Alcohols as Fuel, U.S. Senate, Washington,
     D.C. (May 16, 1974).
158. T. B. Reed, R. M. Lerner, E. D. Hinkley, and R. E. Fahey,
     "Improved Performance of Internal Combustion Engines Using 5-30%
     Methanol in Gasoline," Lincoln Lab. and Energy Lab., Massachusetts
     Inst. of Technol., Cambridge, Mass. (1974).
159. P. Breisacher and R. Nichols, "Fuel Modification: Methanol Instead
     of Lead as the Octane Booster for Gasoline," Meeting of the Combus-
     tion Inst., Madison, Wisc. (Mar. 26-27, 1974).
160. J. C. Ingamells and R. H. Lindquist, "Methanol as a Motor Fuel,"
     Chevron Research Co., Richmond, Calif. (1974).
161. J. M. Colucci, "Methanol Gasoline Blends: Automotive Manufacturer's
     Viewpoint," General Motors Research Labs., presented at the 1974
     Engineering Foundation Conf., Henniker, N.H. (July 7-12, 1974).
162. E. S. Starkman, R. F. Sawyer, R. Carr, G. Johnson, and L. Muzio,
     "Alternative Fuels for Control of Engine Emissions," J.A.P.C.A.,
     Vol. 20 (1970).
163. G. D. Ebersole, "Power, Fuel Consumption, and Exhaust Emission,
     Characteristics of an Internal Combustion Engine Using Isooctane
     and Methanol," Ph.D. thesis, Univ. of Tulsa, Tulsa, Okla. (1971).
164. H. G. Adelman, D. G. Andrews, and R. S. Deboto, "Exhaust Emis-
     sions from a Methanol-Fueled Automobile," Soc. Automotive Engrs.,
     National West Coast Meeting, San Francisco, Calif. (Aug. 21-24,
     1972).
165. W. Bernhardt, W. Lee, and A. Konig, "Methanol," Executive Sum-
     mary, Vol. 1, Research Div., Volkswagenwerk AG, Wolfsburg,
     West Germany (Oct. 1976).
166. B. Baratz, R. Ouellette, W. Park, and B. Stokes, "Survey of Alcohol
     Fuel Technology," Vol. 1, MITRE Rept. M-7-4-61, The Mitre Corp.
     (Nov. 1975).
167. E. S. Starkman, H. K. Newhall, and R. D. Sutton, "Comparative
     Performance of Alcohol and Hydrocarbon Fuels," in Alcohols and
     Hydrocarbons as Motor Fuels, SP-254, Soc. Automotive Engrs.,
     New York (1964).
168. R. W. Duhl and T. O. Wentworth, "Methyl Fuel from Remote Gas
     Sources," Vulcan-Cincinnati, Inc. (April 16, 1974).
169. P. M. Jarvis, "Methanol as Gas Turbine Fuel," General Electric
     Gas Turbine Products Div., paper presented at 1974 Eng. Found.
     Conf., Henniker, N.H. (July 7-12, 1974).

170. R. M. McGhee, "Methanol Fuel from Natural Gas," paper presented at 1974 Eng. Found. Conf., Henniker, N.H. (July 7-12, 1974).

171. J. E. Johnson, "The Storage and Transportation of Synthetic Fuels: A Report to the Synthetic Fuels Panel," Oak Ridge National Lab., Rept. ORNL-TM-4307 (1972).

172. A. Porteous, "A Profitable Means of Municipal Refuse Disposal," Thayer School of Engineering, Dartmouth Coll., Hanover, N.H. (Nov. 1966).

173. A. O. Converse, H. E. Grethlein, S. Karandikar, and S. Kuhrtz, "Acid Hydrolysis of Cellulose in Refuse to Sugar and Its Fermentation to Alcohol," Thayer School of Eng., Dartmouth Coll., Hanover, N.H., NTIS No. PB-221 239 (June 1973).

174. H. E. Grethlein, "The Acid Hydrolysis of Refuse," presented at the NSF Special Seminar on Cellulose as a Chemical and Energy Resource, Univ. of California, Berkeley (June 25-27, 1974).

175. R. A. Chapman, "Acid Hydrolysis of Cellulose in Municipal Refuse," Rept. RC-02-68-11, Public Health Service, U.S. Dept. of Health, Education, and Welfare (1970).

176. G. Mitra and C. R. Wilke, "Continuous Cellulose Production," paper presented at the NSF Special Seminar on Cellulose as a Chemical and Energy Resource, Univ. of California, Berkeley (June 25-27, 1974).

177. C. R. Wilke and G. Mitra, "Process Development Studies on the Enzymatic Hydrolysis of Cellulose," paper presented at the NSF Special Seminar on Cellulose as a Chemical and Energy Resource, Univ. of California, Berkeley (June 25-27, 1974).

178. D. L. Miller, "Industrial Alcohol from Wheat," paper presented at the 6th National Conf. on Wheat Utilization Research, Oakland, Calif. (Nov. 5-7, 1969).

179. D. L. Miller, "Fuel Alcohol from Wheat," paper presented at the 7th National Conf. on Wheat Utilization Research, Manhattan, Kans. (Nov. 3-5, 1971).

180. D. L. Miller, "Agricultural and Industrial Energy," paper presented at the 8th National Conf. on Wheat Utilization Research, Denver, Col. (Oct. 10-12, 1973).

181. S. J. W. Pleeth, Alcohol: A Fuel for Internal Combustion Engines, Chapman & Hall, London, 1949.

182. "Use of Alcohol in Motor Gasoline: A Review," American Petroleum Inst., Committee for Air and Water Conservation, Publ. No. 4082 (Aug. 1971).

183. A. J. Bryce, "A Research and Development Program to Assess the Technical and Economic Feasibility of Methane Production from Giant Brown Kelp," paper presented at the 9th Synthetic Pipeline Gas Symposium, Chicago, Ill. (Oct. 31-Nov. 2, 1977).

184. W. J. North, "Biological Studies of M. pyrifera Growth in Upwelled Water," California Inst. of Technol., Pasadena, Calif., ERDA/USN/ 1027/76/4 (1975).

185. A. Mitsui, "A Survey of Hydrogen Producing Photosynthetic Organism in Tropical and Subtropical Marine Environments," Rosenstiel School of Marine and Atmospheric Sciences, Univ. of Miami, Miami, Fla. (April 1976).

186. A. Mitsui, "Bioconversion of Solar Energy in Salt Water: Photosynthetic Hydrogen Production Systems," Rosenstiel School of Marine and Atmospheric Science, Univ. of Miami, Miami, Fla. (1975).

187. R. C. Landis, "A Technology Assessment Methodology: Mariculture (Sea Farming)," MTR-6009, Vol. 5, The Mitre Corp., Washington, D.C. (June 1971).

Index